Collaborative Genomics Projects: A Comprehensive Guide

Collaborative Genomics Projects: A Comprehensive Guide

Margi Sheth

Human Genome Sequencing Center, Baylor College of Medicine, Houston, TX, United States; The Cancer Genome Atlas, Center for Cancer Genomics, National Cancer Institute, National Institutes of Health, Bethesda, MD, United States

Jiashan Zhang

The Cancer Genome Atlas, Center for Cancer Genomics, National Cancer Institute, National Institutes of Health, Bethesda, MD, United States

Jean C. Zenklusen

The Cancer Genome Atlas, Center for Cancer Genomics, National Cancer Institute, National Institutes of Health, Bethesda, MD, United States

AMSTERDAM • BOSTON • HEIDELBERG • LONDON
NEW YORK • OXFORD • PARIS • SAN DIEGO
SAN FRANCISCO • SINGAPORE • SYDNEY • TOKYO

Academic Press is an imprint of Elsevier

Academic Press is an imprint of Elsevier
125 London Wall, London EC2Y 5AS, UK
525 B Street, Suite 1800, San Diego, CA 92101-4495, USA
50 Hampshire Street, 5th Floor, Cambridge, MA 02139, USA
The Boulevard, Langford Lane, Kidlington, Oxford OX5 1GB, UK

First Edition 2016

Notices
Knowledge and best practice in this field are constantly changing. As new research and experience broaden
our understanding, changes in research methods, professional practices, or medical treatment may become
necessary.

Practitioners and researchers must always rely on their own experience and knowledge in evaluating and
using any information, methods, compounds, or experiments described herein. In using such information or
methods they should be mindful of their own safety and the safety of others, including parties for whom
they have a professional responsibility.

To the fullest extent of the law, neither the Publisher nor the authors, contributors, or editors, assume any
liability for any injury and/or damage to persons or property as a matter of products liability, negligence
or otherwise, or from any use or operation of any methods, products, instructions, or ideas contained in
the material herein.

ISBN: 978-0-12-802143-9

British Library Cataloguing-in-Publication Data
A catalogue record for this book is available from the British Library.

Library of Congress Cataloging-in-Publication Data
A catalog record for this book is available from the Library of Congress.

For Information on all Academic Press publications
visit our website at http://store.elsevier.com/

Typeset by MPS Limited, Chennai, India
www.adi-mps.com

Printed and bound in the USA

Contents

CHAPTER 1 **Introduction** 1
Overview of the Cancer Genome Atlas 1
Scope, Implementation, and Applicability of This Guide...................... 3
Policy Considerations 4
References.. 5

CHAPTER 2 **Gathering Project Requirements** 7
Introduction.. 7
Establish the Purpose of the Project 7
Identify Key Stakeholders 9
Set Project Milestones 9
Design a Pipeline for the Project 10
    Examples of Pipeline Activities... 13
Conclusion 14
References.. 14

CHAPTER 3 **Communications Strategies** 15
With contribution by Emma Spaulding, Center for Cancer Genomics, National Cancer Institute, National Institutes of Health

Introduction 15
Why Develop a Communications Strategy? 15
    A Note on Press Releases... 16
    A Communications Strategy Applies Tactics to Share Messages....... 16
How to Develop a Communications Strategy 17
    Step 1: Audiences: Who Are the Audiences and Stakeholders?........ 17
    Step 2: Challenges: What Are Their Concerns?........................... 18
    Step 3: Consideration: Are the Audience's Concerns Valid? 19
    Step 4: Messaging: What Should Be Said in Response? 20
    Step 5: Tactics: What Is the Best Way to Reach the Audience?........ 21
    Step 6: Priority: Which Audiences and Messages Are Most Important?... 21
    Step 7: Tactic Evaluation: Were the Communications Messages Received? 22
    Step 8: Strategy Evaluation: Are the Tactics Working? 22
Project and Policy Changes: A Part of Strategy Evaluation 23
Examples from Communications Strategies 23
Communications Devices: Symposia and Visual Identity..................... 26
    Annual Scientific Symposia 26
    Visual Identity 27

Conclusion 28
References... 29

CHAPTER 4 Pipeline: Sample Acquisition 31
Introduction.. 31
Define the Sample Set for the Project 31
Establish a Central Biospecimen Processing Facility............................. 32
Establish Sample Qualification Metrics.. 33
Sample Processing and Distribution to Data Generation Centers............ 34
Establish Consent Protocols 35
Handling Institutional Review Boards in Multi-Center Studies............... 36
Implications of Using a Centralized IRB.............................. 36
Implications of Honoring Individual IRB Approval Rulings.............. 36
Identify Potential Tissue Source Sites 37
Establish Contractual Obligation and Payment Plans
for Tissue Source Sites.. 37
Management of Clinical Data Collection 38
Sharing Clinical Data 40
Practical Considerations for Clinical Data Collection
and Management... 40
References... 41

CHAPTER 5 Pipeline: Data Generation 43
Introduction.. 43
Building a Data Generation Model 43
What Data Types Should Be Generated? 43
What Technologies and Methods Should Be Employed
to Generate Data? 45
Who Will Be Generating the Data? 45
Establishing a Data Generation Pipeline and Quality Control
Measures 47
Proper Tracking of Data Generation... 48
Conclusion 48
References... 49

CHAPTER 6 Pipeline: Data Storage and Dissemination 51
Introduction.. 51
Creation of Centralized Data Management Center.......................... 51
Define Standard Data and Metadata Formats 52
Collect, Store, and Version Data and Metadata 53
Implement Quality Control Measures for Submitted Data................. 55
Put in Place Appropriate Security and Access Controls 56

		Redistribute Data and Metadata Tailored to Diverse Project Stakeholders and End Users	56
		Conclusion	57
		References	58
CHAPTER 7	**Pipeline: Data Analysis**		**59**
		Introduction	59
		Preconceived Questions to Answer	59
		Establishment of Data Analysis Teams	61
		Analysis Teams and Diversity of Expertise	61
		Managing Contributions	61
		Analysis Structure and Methodology	62
		Study Design	62
		Emerging Analytical Tools	63
		Outliers/Exceptional Cases	63
		Practical Considerations	64
		Falling Into a "Formula"	64
		Aiming for Journals	64
		Authorship Models	65
		Timeliness Versus Scientific Merit	65
		Conclusion	66
		References	67
CHAPTER 8	**Quality Control, Auditing, and Reporting**		**69**
		Introduction	69
		Establish Quality Metrics for Each Component of the Pipeline	69
		Ensure Ethical Management of Samples, Information, and Derived Data Sets	70
		Provide Quality Reports to Stakeholders to Help Improve Processes	71
		Example of a Quality Management Issue	71
		Conclusion	72
CHAPTER 9	**Project Closure**		**73**
		Introduction	73
		Levels of Closure	73
		Publication Level	73
		Institutional Level	73
		Program Level	77
		Budgetary Considerations	77
		Budgetary Vigilance by Program Office	78
		Budgetary Vigilance by Data Providers	78

Conclusion 78
References 79

Conclusion **81**
Flexibility 81
Transparency 82
Collaboration 82
Communication 82
References 83

Appendix A: Glossary of Terms 85
Appendix B: TCGA Workflow Diagrams 91
Appendix C: Publication Guidelines as of July 14, 2015 97
Appendix D: Mutation Annotation Format (MAF) Specification 103
Appendix E: MAGE-TAB 111
Appendix F: Data Use Certification Agreement 115
Appendix G: TCGA Analysis Working Group Charter 127
Index 133

Chapter

1

Introduction

As the cost of genomic sequencing is decreasing, more and more research-ers are leveraging genomic data to inform the biology of disease. The amount of genomic data generated is growing exponentially, and protocols need to be established for the long-term storage, dissemination, and regula-tion of these data for research. We aim to create a comprehensive guide to managing research projects involving genomic data, as learned through the evolution of the Cancer Genome Atlas (TCGA) program over the last decade. This project was primarily carried out in the United States, but the impact and lessons learned can be applied to an international audience.

The guide will serve to:

- Establish a framework for managing large-scale genomic research projects involving multiple collaborators,
- Describe lessons learned through TCGA to prepare for potential roadblocks,
- Evaluate policy considerations that are needed to avoid pitfalls, and
- Recommend strategies to make project management more efficient.

The guide will cover operational procedures, policy considerations, and lessons learned through TCGA on the following topics:

- Sample acquisition
- Data generation
- Data storage and dissemination
- Data analysis
- Quality control, auditing, and reporting
- Project closure
- Communications

OVERVIEW OF THE CANCER GENOME ATLAS

In 2006, the National Cancer Institute (NCI) and the National Human Genome Research Institute (NHGRI) initiated a pilot project to determine the feasibility of comprehensively cataloging the genomic alterations

Collaborative Genomics Projects: A Comprehensive Guide. DOI: http://dx.doi.org/10.1016/B978-0-12-802143-9.00001-4

associated with three different human cancers. This initial pilot project demonstrated that cancer-associated genes and genomic regions can be identified by combining diverse genomic information with tumor biology and clinical data, and that the sequencing of selected regions can be conducted efficiently and cost-effectively. In 2009, TCGA expanded to characterizing 33 different types of human cancers including nine rare cancers. The strength of TCGA was producing unprecedented multidimensional data sets using an appropriate number of samples to provide statistically robust results.

Three overarching lessons were learned through TCGA in order to successfully interpret results generated by various genomic characterization platforms. Data generation centers had to (1) utilize high-quality molecular analytes isolated from well-characterized tissue specimens, (2) perform experiments utilizing strictly standardized protocols, and (3) deposit the results in structured and well-described formats. The last lesson strongly impacted on the ability of the various analytical groups to extract meaningful results from the genomic data generated.

The unique aspect of TCGA was the development and function of an integrated research network. The intent of TCGA was to conduct a coordinated, comprehensive, genome-wide analysis of cancer-relevant alterations by simultaneously applying several technologies to interrogate the genome, epigenome, and transcriptome in large collections of quality-controlled cancer biospecimens derived from specific cancer types. To accomplish this goal, TCGA included multidisciplinary teams of investigators and associated institutions that collectively provided biological data, as well as informed strategies for data analysis through the development of bioinformatics tools. The progress in understanding some cancer-associated molecular alterations and the accompanying advances in technology suggested that it was possible to obtain comprehensive genomic information from multiple tumor types to catalog most, if not all, of the genomic changes associated with cancer. The TCGA Research Network demonstrated that a coordinated pipeline approach for the investigation of cancer is the best way to avoid biases in the data sets, thus allowing for interoperability of the different cancer type–specific analysis projects.

TCGA followed a coordinated pipeline to receive tissues accrued, process analytes, generate and analyze data, and present the results to the community. Components of this pipeline were:

■ Biospecimen Core Resource (BCR): The BCR served as the tissue processing center and provided the analytes for Genome Characterization Centers (GCCs) and Genome Sequencing Centers (GSCs). Standard operating procedures were used for clinical data collection, sample collection, pathological examination, analyte (eg, DNA and RNA) extractions, quality control, laboratory data collection, and analyte distribution to the GCCs and GSCs. The samples were required to have patient-informed consent for the public release of data or an IRB waiver.

- Genome Characterization Centers (GCCs) and Genome Sequencing Centers (GSCs): The GCCs and GSCs produced high-quality genomic, transcriptomic, proteomic, and epigenomic data using validated technologies (eg, DNA and RNA sequencing, methylation arrays, etc.) to reveal the spectrum of alterations that exist in human tumors.
- Genome Data Analysis Centers (GDACs): The GDACs worked hand-in-hand with the GCCs and GSCs to perform "higher level" analyses of the data produced by the GCCs and GSCs and to develop state-of-the-art tools that assist researchers with processing and integrating data analyses across the entire genome. These analyses took the form of both fully automated pipelines, as well as ad hoc analyses performed at the request of the Analysis Working Groups (AWG) from each project.
- Data Coordination Center (DCC) and Cancer Genomics Hub (CGHub): Data generated by GCCs and GSCs were deposited into central repositories—the DCC and CGHub, as soon as they were validated, in general within a few weeks of generation. CGHub handled raw sequence data generated from the GCCs and GSCs, while the DCC handled higher level interpreted data. Data submitted to each repository followed rigorously developed data standards and controlled vocabularies for new technology platforms. Both DCC and CGhub also created portals for basic and clinical researchers to easily access the data, as well as provided a secure network and means to protect the integrity and security of genomic and clinical data.
- Analysis Working Groups (AWG). As mentioned above, TCGA formed multidisciplinary teams of investigators, disease experts, bioinformaticists, etc. to analyze, interpret, and correlate the array of genomic, transcriptomic, proteomic, and epigenomic data generated for each cancer type–specific sample set. These teams ultimately produced a marker publication that included novel findings, clinical correlations, and if relevant, therapeutic implications for a particular cancer type. These publications and associated data sets were extremely valuable to the research community.

SCOPE, IMPLEMENTATION, AND APPLICABILITY OF THIS GUIDE

This guide aims to provide practical considerations for managing large-scale genomics research projects and is by no means a step-by-step instruction book. Due to the diversity of genomics research projects, it is nearly impossible for all projects to fit one management model. The authors make certain assumptions when discussing large-scale genomics research projects:

- The project involves human subjects.
- The project is multi-institutional (ie, multiple sample providers).
- The project involves genomic data generation.
- The project is large-scale, that is, sample sets are obtained from a variety of populations.
- The project involves many substudies on different populations.

While this guide features lessons learned from TCGA, we also include examples from other large-scale genomics research projects funded by the National Institutes of Health (NIH). The suggestions in this guide are merely recommendations; they are not a foolproof way to successfully manage a large-scale genomics research project.

POLICY CONSIDERATIONS

With research projects involving human subjects, one must ensure that protocols follow all ethics rules and regulations. These can be categorized broadly into human subject protection and data sharing policies. Principles from The Common Rule [1] and the Human Subjects Research Policy (45 CFR 46) [2] must be readily apparent in all research activities.

Below are overview of policy considerations for each component of the project pipeline (discussed in detail in Chapter 2: Gathering Project Requirements).

1. Sample acquisition and characterization: Human subjects protection legislation must be followed when defining goals related to number of patients, information gathered from patients, and intricacies of patient consent. Data sharing protocols between contributing sites must not violate laws of each participating nation in the project.
2. Data generation: Data sharing within institutions during data generation (ie, Laboratory Information Management Systems) must follow the project's data sharing policies.
3. Data storage and dissemination: This process requires heavy scrutiny to ensure patient privacy and confidentiality are protected. Clear definitions of publicly accessible and protected data sets must be drawn [3].

4. Data analysis efforts: Automated tools developed for analysis must have the appropriate privacy and confidentiality controls built in for data sharing, especially if the tools are made for public use. If data analysis suggests further data acquisition can enhance findings, patient privacy policy and established contractual obligations must be respected.

5. Quality control, auditing, and reporting: Quality control protocols, especially those of data sharing and dissemination, should include built-in checkpoints to ensure the activity complies with patient privacy and data sharing policies. Auditing and reporting practices should explain violations, corrective actions, and consequences for policy breach.

REFERENCES

[1] Code of federal regulations. Retrieved from HHS: <http://www.hhs.gov/ohrp/humansubjects/guidance/45cfr46.html>; October 26, 2015.

[2] Federal policy for the protection of human subjects (common rule). Retrieved from HHS: <http://www.hhs.gov/ohrp/humansubjects/commonrule/index.html>; October 26, 2015.

[3] Genomic data sharing policy. Retrieved from NIH: <https://gds.nih.gov/03policy2.html>; October 26, 2015.

Chapter 2

Gathering Project Requirements

INTRODUCTION

When starting a large-scale project involving many components, the first task is to gather all project requirements. Who are the stakeholders? How many subjects will the study include? What human subjects protection policies need to be implemented? What kinds of subject-related metadata are needed? How much genomic data will be generated? What infrastructure is needed to properly store, version, standardize, and distribute data sets to the user community? These are just a sampling of the types of issues that will need to be addressed before embarking on a large-scale genomics research project. In the following sections, we provide an overview of how to establish a standardized pipeline for conducting such a project. In-depth discussion of pipeline components will be elaborated in later chapters of this book.

ESTABLISH THE PURPOSE OF THE PROJECT

The first and foremost element to establish before embarking on a genomics research project is a *purpose*. A purpose statement should be written in unambiguous terms and shared with all participants of the project. This enables participants to work toward a single objective. The purpose statement should be able to answer the following questions:

1. What is the goal of the project?
2. How will the goal be accomplished?
3. Whom does the project serve?
4. Why is the project significant or valuable?

Collaborative Genomics Projects: A Comprehensive Guide. DOI: http://dx.doi.org/10.1016/B978-0-12-802143-9.00002-6

Below is an example of a clear purpose statement from the 1000 Genomes Project [1].

> *The aim of the 1000 Genomes Project is to discover, genotype and provide accurate haplotype information on all forms of human DNA polymorphism in multiple human populations. Specifically, the goal is to characterize over 95% of variants that are in genomic regions accessible to current high-throughput sequencing technologies and that have allele frequency of 1% or higher (the classical definition of polymorphism) in each of five major population groups (populations in or with ancestry from Europe, East Asia, South Asia, West Africa, and the Americas). The 1000 Genomes Project is the first project to sequence the genomes of a large number of people to provide a comprehensive resource on human genetic variation. As with other major human genome reference projects, data from the 1000 Genomes Project will be made available quickly to the worldwide scientific community through freely accessible public databases.*

From the above purpose statement, clear answers to each of the questions can be inferred:

1. What is the goal of project?

   The aim of the 1000 Genomes Project is to discover, genotype and provide accurate haplotype information on all forms of human DNA polymorphism in multiple human populations.

2. How will the goal be accomplished?

   ...characterize over 95% of variants that are in genomic regions accessible to current high-throughput sequencing technologies and that have allele frequency of 1% or higher (the classical definition of polymorphism) in each of five major population groups (populations in or with ancestry from Europe, East Asia, South Asia, West Africa and the Americas).

3. Whom does the project serve?

   ...the worldwide scientific community...

4. Why is the project significant or valuable?

   The 1000 Genomes Project is the first project to sequence the genomes of a large number of people, to provide a comprehensive resource on human genetic variation.

IDENTIFY KEY STAKEHOLDERS

By definition, a *stakeholder* is an individual, group, or organization that is invested in a project by affecting the project or being affected by the results of project. For large-scale genomics research projects, stakeholders may include:

- Parties involved in project funding: government funding agencies, nonprofit funding organizations, corporate sponsors, private donors, etc.
- Parties involved in project operations: management staff, research staff, contract organizations, etc.
- Parties affected or served by the project: research community, patients, patient advocates, etc.

Appropriate involvement of stakeholders at project inception is vital. For example, funding institutions provide the initial monetary investment into a project and determine whether the project will continue to be funded in the next phases. Funders should be kept informed of project progress via regular reports (see chapter 8: Quality Control, Auditing, and Reporting). The research community served by the project should be asked to provide feedback on useful data types pertaining to their research.

A strategy should be developed for communicating project results to each of the key stakeholders in a typical genomics research project—funding agencies, collaborators, scientific community, and patients or patient advocates (for projects aimed at improving health outcomes). These are different groups that have a unique understanding or interest in the project. See chapter 3: Communication Strategies for developing communication strategies for different types of audiences.

SET PROJECT MILESTONES

It is important to develop a project plan with concrete milestones early. A milestone should be defined as completion of a certain task in the plan (or set of tasks). Milestones should be practical, realistic, and agreed upon by all participants of the project. Milestones are not measured in time, but rather in the achievement of steps toward the project goal.

For large-scale genomics research projects, the first milestone should be designing the project pipeline (see next section: Design a Pipeline for

Table 2.1 General Milestones for Large-Scale Genomics Research Projects

Milestone 1	Design project pipeline (and protocols for each pipeline activity)
Milestone 2	Set up infrastructure for each pipeline activity
Milestone 3	Complete sample acquisition and define sample set
Milestone 4	Complete data generation for sample set
Milestone 5	Submission of all data to data repository
Milestone 6	Complete analysis of all data (eg, prepare publications)
Milestone 7	Disseminate data to research community as appropriate

the Project). The next milestone should be the completion of infrastructure setup for the project pipeline. This involves identifying the appropriate setting for each major project activity and gathering the necessary human and material resources. Once the pipeline is established, subsequent milestones may include completion of sample set acquisition, completion of data generation and analysis, completion of data submission, construction of data repository, delivery of an end product, or other related achievements. The number of milestones depends on the complexity of the project. Table 2.1 shows a sample milestone chart for a genomics research project. The TCGA program informally followed these milestones.

It is important to note that project activities may overlap. For example, data generation and analysis may occur before all samples have been acquired. In-depth discussion of best practices for setting up each component of the pipeline will be discussed in later chapters.

DESIGN A PIPELINE FOR THE PROJECT

When designing a large-scale genomics research project, it is essential to determine project activities. Once project activities are defined and their relationships established, a pipeline diagram should be created to elucidate a sequential flow of tasks.

1. Establish project activities: Project activities are discrete steps required to achieve the project goal. The activities may be performed by different collaborator sites according to area of expertise. These activities should be derived from the purpose statement discussed above.

Each activity should have clearly stated goals, deliverables, human and monetary resource allocation, and timeline for completion. The activities serve as the basis for defining pieces of the project pipeline. Refer to next section: Build the Pipeline Diagram, for example, of project activities.

2. Establish relationships between activities: The interactions between discrete activities should be defined. Can a completed step be revisited downstream if it seems beneficial to do so based upon a new finding? Should relationships between activities be two-way to allow for back-and-forth processes that may arise when having difficulties reaching a goal? What items are final and unchangeable once completed? Questions such as these must be tackled when designing the interaction scheme between activities and be approached with flexibility when they occur. Gantt charts, flowcharts, and other visual organizational tools can greatly help clarify relationships.

   Accountability tracking within and between activities is critical when designing a pipeline with large interacting components. Problems will inevitably arise due to miscommunication, human error, or machine error. Tracking of all actions, with names and dates applied, can help trace the origin of error. A clear handoff policy can define the transfer of responsibility from one party to another. Especially in tasks with high detail and sensitivity (such as lab protocol and data coordination), it is helpful to define the accountable party(ies) for each step in the process. This also allows for easier troubleshooting for an error or problem that may occur within the process. See chapter 8: Quality Control, Auditing, and Reporting for more information.

   Activities may be dependent upon each other in varying capacities. Designating criteria of the activity's end product (ie, deliverable) is a finer detail which may be necessary at this early stage to help define relationships. Criteria may include qualification standards, quality metrics of deliverables, sample size goals, or analytical outcomes of biological data. For example, data generated by the project cannot be submitted to the data repository without conforming to data standards set by the project.

3. Build the pipeline diagram: This is most easily represented by a flow chart. The pipeline diagram should clearly show the interdependency of each project activity. Refer to the TCGA workflow diagrams in Fig. 2.1 and Appendix B.

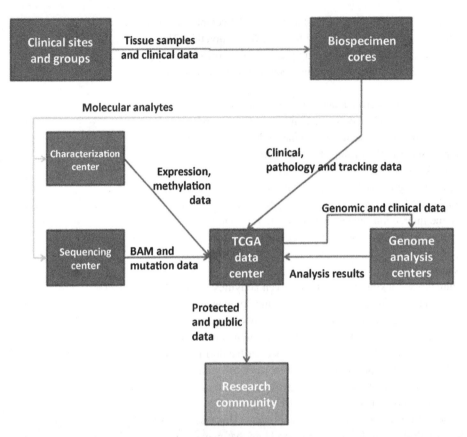

■ **FIGURE 2.1 TCGA Workflow Diagram.** TCGA created a community data resource by collecting samples and building a cancer genomics data set. Tissue banks and cancer clinics comprised the Tissue Source Sites (TSS). They provide tissue samples and clinical data to centralized sample processing sites called Biospecimen Cores (BCRs). The BCRs reviewed the sample quality and determined whether the sample met inclusion criteria before enrollment into TCGA. Upon enrollment, the BCRs sent clinical data to the TCGA Data Coordinating Center (DCC) and apportioned molecular samples for disbursement to the Genome Characterization and Sequencing Centers (GCCs and GSCs). Note that the GCCs, GSCs, and BCRs independently send data and metadata to the DCC and Cancer Genomics Hub (CGHub). The DCC and CGHub collects and coordinates all public and protected data, synthesizing it for distribution to the research community. The Genome Data Analysis Centers (GDACs) have a two-way arrow with the DCC; they obtain characterization/sequencing data through the DCC, analyze the data, and ultimately store their analysis results at the DCC.

This was an original workflow diagram publicized on the TCGA website as the study began. The shortcoming of this diagram is that it does not show the interactions between the Characterization, Sequencing, and Analysis Centers. If the flowchart were to expand upon the data analysis process, the BCRs, Characterization, and Sequencing centers would have two-way arrows with the Data Coordination Center. This would depict the constant intellectual exchange between the centers that result in additional data generation and re-analysis when appropriate. Updated workflow figures for the TCGA program can be viewed in Appendix B: TCGA Workflow Diagrams.

Examples of Pipeline Activities

In the following section, we list the typical activities involved in a large-scale genomics research project and considerations needed to define each activity. Background research and specific expertise may be required to establish standards and quality metrics for each activity.

- **Sample acquisition and enrollment**
  - Quantity of samples needed to generate desired data
  - Biomedical and sample inclusion criteria and evaluation
  - Location of sample collection and storage (ie, biorepository)
  - Frequency and threshold of quality control checks per sample
  - Patient consent requirements necessary for data generation, analysis, and sharing
  - Sample aliquot disbursement and shipping protocol
  - If samples are obtained from multiple sources, additional considerations include the following:
    - reimbursement policy for obtaining outside samples and clinical data
    - contract establishment and execution for sample acquisition
    - training and education of sample providers regarding project participation
- **Data generation**
  - Determining the specific technologies needed to generate data
  - Laboratory protocols for each technology
  - Data quality metrics
  - Accounting of samples through data generation pipeline
  - Protocols for cleaning and curating raw data
- **Data storage and dissemination**
  - Protocols for standardizing data formats
  - Methods for versioning data and collecting metadata
  - Patient consent obtained for analysis and sharing of data internally and externally
  - Considerations for public and protected access data sets
  - Short term and long-term data storage and archive plans
  - Methods for disseminating data to user community (ie, user interfaces)
- **Data analysis efforts**
  - Establishment of standard, automated analyses per sample/sample set
  - Development of analysis tools
  - Establishment of collaborative working groups to analyze data

- **Quality control, auditing, and reporting**
  - ❑ Quality control metrics for each component of the pipeline
  - ❑ Quality control feedback system
    - Regular audits of all data produced by the project
    - Reporting tools to engage key stakeholders

Interplay between activities: Resolution of sample swaps that occurred in TCGA illustrates the interplay of activities and utility of detailed accountability tracking. Suppose a small subset, perhaps 5% of the sample set, displays abnormal results on a single data platform such as mRNA sequencing. This sample subset is analyzed for novel genetic and/or epigenetic expression, but a clear answer does not present itself. Comparison with SNP data from the same subset to verify genotype and expression concordance fails, suggesting sample identities do not match [2]. Which samples were mislabeled, and where did the swap occur? The clear allocation of responsibilities to each activity participant allows one to follow these samples through each step. Tracking these samples through sample collection, quality control, aliquoting, shipping, arrival quality control, data generation, and analysis can help pinpoint where and when a sample swap may have occurred. Scientists from each activity work together to resolve the issue, but lines of accountability are clearly drawn. Once the error is identified, the team may decide to re-characterize these samples. A clear pipeline workflow statement or diagram would describe how to redo this procedure at the necessary step with sample identity intact (if resources are sufficient).

■ CONCLUSION

Developing a clear purpose statement, identifying key stakeholders, developing a project plan, defining milestones, and establishing a workflow diagram for the project pipeline is essential for proper project planning. For genomics research projects, the pipeline should generally have the following components: sample acquisition, data generation, data storage and dissemination, data analysis, and quality control and reporting systems. The interplay between components should be carefully laid out and adjusted as necessary.

REFERENCES

[1] 1000 Genomes: A deep catalog of human variation, <http://www.1000genomes.org/about> [cited December 12, 2014].
[2] Huang J, Chen J, Lathrop M, Liang L. A tool for RNA sequencing sample identity check. Bioinformatics 2013;29(11):1463−4.

Chapter 3

Communications Strategies

With contribution by Emma Spaulding, Center for Cancer Genomics, National Cancer Institute, National Institutes of Health

INTRODUCTION

As Lee Atwater, Former President George H. W. Bush's 1988 campaign advisor, said, "Perception is reality." If no one finds out about the project, or consequently uses the data, were the project goals achieved? Communications are owed to the stakeholders of the project, particularly if the project is government backed and funded by the taxpayers.

Communications means more than "getting the word out" or publicizing a journal article. Communications is a way to inform interested parties about the project. A well thought-out communications plan will strategically disseminate relevant information about the project. This chapter outlines eight steps to develop a communications strategy and examples of each step.

WHY DEVELOP A COMMUNICATIONS STRATEGY?

A communications strategy is designed to share key messages with identified audiences. A clear and insightful communications strategy advances project goals.

> *A communications strategy unifies messages from project leaders who may come from different institutions and even different countries.* A large collaborative project necessarily brings together experts from several specialties from different institutions. As prominent figures in the project, each expert will have the opportunity to communicate about the project, whether in a press release from their home institution or an interview on television. Dissonant messages from these experts may undermine the project by making its goals seem ambiguous and underdeveloped. However, consistent communications messages show that the project is united to achieve its goals.

Collaborative Genomics Projects: A Comprehensive Guide. DOI: http://dx.doi.org/10.1016/B978-0-12-802143-9.00003-8

A communications strategy positions the project to take control of what's said about it. A large collaborative project will garner attention from the scientific community and other audiences as well. People will discuss the project with or without efforts implemented by a communications strategy. By contributing to the conversation, it's possible to shape the discussion and influence opinions about the project.

With a thoughtful communications strategy, the value of a well-structured project working towards an important goal will be clear. Science should not stand unsupported. Communicating exclusively through scientific articles in peer-reviewed journals will leave out many audiences or will not provide sufficient context for some of the audience members the articles do reach. When the scientific goals of a project are made clear through targeted communications, the value of a well-developed project should become apparent.

A Note on Press Releases

Press releases are *part* of a communications strategy—they should not *be* the communications strategy. Reporters are bombarded with press releases and story pitches all day, while also trying to write on a deadline. Given a reporter's limited time, the odds of a single press release becoming a feature story are slim. Press releases are most effective as a component of a communications strategy, either supplemented by a professional relationship with a reporter or as part of a greater media and communications push from the home institution.

Press releases can be complex to draft and distribute, especially for large international projects. Work with the home institution's media office to develop a plan for review. A flowchart for review and clearance of the press release by subject matter experts ensures that feedback is gathered from required parties and that the release is accurate.

A Communications Strategy Applies Tactics to Share Messages

A communications strategy outlines the overarching communications messages and objectives for a project. Tactics are the means to achieve an objective. The strategy is the "What?" of the project (What is an important aspect of the project?), and the tactics are the "How?" (How will this be communicated?). Before determining the tactic, a strategy and the relevant messages must be formed.

A common mistake in creating communications pieces is to choose a tactic and then decide the message. An example of putting a tactic before the message would be deciding to make a video for YouTube, and then determining that the topic should be a tutorial of how to access data. Here, the tactic (a YouTube video) has been chosen before the message ("Resources are available to assist in accessing data").

Perhaps the audience for the data access tutorial is too impatient enough for a video, or the steps in the video move too quickly. Perhaps this is the organization's first YouTube video, which would necessitate promotion of a new form of content (not to mention creating and getting approval for an official institutional YouTube account). A better method would be to determine the communications strategy and message, which is to explain data access, and then determine the best tactic. The best tactic may be a flowchart outlining the steps to access the data so that a user can work through those steps at his or her own pace. By considering strategy and message before tactic, better and simpler mediums to communicate the message may be identified.

HOW TO DEVELOP A COMMUNICATIONS STRATEGY

Use this eight-step process to develop the strategy.

1. Audiences: Who are the audiences and stakeholders?
2. Challenges: What are their concerns?
3. Consideration: Are the audience's concerns valid?
4. Messaging: What should be said in response?
5. Tactics: What is the best way to reach the audience?
6. Priority: Which audiences and messages are most important?
7. Tactic Evaluation: Were the communications messages received?
8. Strategy Evaluation: Are the tactics working?

Step 1: Audiences: Who Are the Audiences and Stakeholders?

The audiences are categories of people who will either be interacting with or will want know about the project. Stakeholders, as previously defined, are groups of people who are invested in the project. The stakeholders within each audience may be highlighted to receive special communications efforts.

Listed here are some potential audience categories:

1. Scientists: This group includes basic and clinical researchers, trainees, bioinformaticians, bench scientists, and translational researchers. This

audience could be further broken down into scientists within the organization and scientists outside of the organization.

2. Industry: This group comprises biotechnology companies, pharmaceutical companies, bioinformatics companies, genomic sequencing companies, and genomic data production centers.

3. Health Care Providers: This audience includes doctors, nurses, physician's assistants, and specialists, like oncologists, pathologists, and radiologists.

4. Patients and Advocacy Organizations: This group subsists of cancer patients, their families and caregivers, as well as members of general cancer advocacy groups, such as the American Cancer Society, or members of cancer subtype advocacy groups, like the International Myeloma Foundation.

5. Government Officials and Policy Makers: This includes members of regional or national government and their staffs, as well as policy influencers like lobbyists or activists.

6. Media: This audience encompasses all members of the media, from national newspapers, like *The New York Times* to smaller outfits, like trade and specialty publications.

7. Public: This group represents members of the general public. This group does not track research in particular and for the most part, has little or no knowledge in science or genomics.

Depending on available resources, consider dividing the communications strategy into two phases based on the importance of each audience. For example, Phase I of the communications strategy might focus on Scientists, Industry, Health Care Providers, and Patients and Advocacy Organizations. The audiences outlined for a Phase I approach are a more immediate priority since they will likely interact with the project before the audiences in Phase II, which would include Government Officials and Policy Makers, Media, and the Public.

Step 2: Challenges: What Are Their Concerns?

After identifying audiences relevant to the communications strategy, understand the aims of each group, and how the project might impact their goals. This information can be gathered formally or informally.

A formal approach would be to send surveys to representative members of each audience, followed up with structured interviews. An informal strategy would involve polling contacts from the relevant audiences. In these conversations, the contact should be able to describe their group's goals and articulate possible concerns about the project.

For example:

- Scientists outside the project may see it as using resources that could go toward R01 grants.
- Health care providers might be hesitant to refer patients to contribute to the project because they do not see how it will help their particular patient.
- Government officials might be concerned that the funding for the project has been siphoned off from money and jobs that could go to their state or country.
- Industry may be concerned that they will not be able to patent findings derived from the data or that their findings will not be considered unique or profitable.
- Patients might be upset if their samples are precluded or excluded from the project and might wonder why their samples were not "good enough."

Step 3: Consideration: Are the Audience's Concerns Valid?

After establishing an audience's concerns, it is important to understand if their perceived concerns are valid. In other words, is there a bug in the project or a weak spot in the communications strategy? If an audience's concern is valid, they may have spotted a flaw in the project design. Consider addressing this by changing the project design or adjusting the communications strategy.

For example:

A bioinformatician may be worried about using the data because the samples have not undergone quality control testing.

- If the quality of the samples has not been tested, consider adding that as a step.
- If the quality of the samples has not been tested and there are no plans to do so, explain why.
  - "Sample collection was retrospective, meaning that the tissue may not have been QC'ed at the time of collection or if it was, that information is not currently available."

- However, the audience's concern may not be valid. In that case, they have spotted a flaw in the communications strategy.
- If the quality of the samples has been tested, develop messages that outline and clarify the sample quality control process.

A data user may want additional patient clinical data that were not collected by the project.

- If the particular clinical data element is important to the analysis, such as diagnosis of Barrett's esophagus in esophageal cancer analysis, considering collecting it as part of the project.
- If the particular clinical data element was not collected as part of the study, explain why.
  - "The original pathology reports from the tissue donation sites did not contain Barrett's esophagus information for the patient or the final diagnosis is unclear in the pathology report."
- If the clinical data element was collected, develop messages that outline this and clarify where to access the data.
- If the concern expresses a desire for elements beyond the scope of the project or program, the scope and aims may need to be communicated more clearly to the particular audience.

Step 4: Messaging: What Should Be Said in Response?

To address the audience's concerns, develop key communications messages. These are stand-alone statements that respond to general categories of concern. While they do respond to concerns, these messages should stand independently, not as argumentative or reactionary statements.

For example:

- Challenge: Some pathologists may believe that TCGA's data are of poor quality because the samples are poor quality.
- Messaging: "TCGA works with Tissue Source Sites and the Biospecimen Core Resource to ensure data and samples are of the highest quality."

- Challenge: Scientists outside the project may see it as usurping resources that could go toward R01 grants.
- Messaging: "TCGA funding was appropriated by the US Congress and was not taken from the NIH grants budget."

Step 5: Tactics: What Is the Best Way to Reach the Audience?

As previously stated, tactics are how the communications messages are disseminated. The medium is where these messages are shared. Examples of a medium are a quarterly newsletter, conference calls, or social media. Good tactics make use of the appropriate medium.

Select the medium to serve the message and audience. Share the message where the intended audience is already comfortable. If the message is not served on a medium that the audience uses, they will not receive the message. Identify key communications mediums the audience uses and deliver them the message there.

For example:

- Bioinformatics postdoctoral fellows may already be using Twitter. On that medium, create and maintain a Twitter account to communicate new collaboration opportunities in the project to them.

- International advocacy groups may be comfortable working with each other over teleconferences. Secure a time slot at their next all-hands meeting to orient them to the project.

Step 6: Priority: Which Audiences and Messages Are Most Important?

Unless the project has limitless resources for communication, it is imperative to identify the messages or audiences most important in the long and short term. Determining the most important audiences (as described in Step 1) will allow a large communications strategy to be divided into a multiple phase approach, with the highest priority audiences and messages as part of the first phase.

Working on a mix of long- and short-term projects allows the more complex, long-term projects to be carried forward, interspersed with some short-term "wins." Additionally, there may be a communications tactic that can address several communications messages at once.

For example:

- A project team member might be featured in a video to be distributed to members of U.S. Congress and their staffs as part of its annual budget and appropriations plan. That content may also be highlighted on the institution's website to explain the significance of the research program's findings to advocates and non expert scientists.

- The TCGA website features several pieces that share multiple communications messages in a single article.
  - A story about a researcher who used TCGA data in a grant application reinforces the message that people can use TCGA data in successful grant applications, and that TCGA is open to outsiders and is not a "members only" club.
  - One piece features a bioinformatician who reinforces the message that TCGA is open to outsiders, and also shares a new tool for scientists to analyze TCGA data.
  - Another article features a scientist entrepreneur which shows industry and researchers that findings derived from TCGA data can be commercialized.

Note that prioritizing some tactics inescapably means that others will not be a first concern. Some of these tactics may never be completed or even executed. This is a necessity of a communications strategy. Deciding not to pursue or start a tactic is as important a decision as choosing to launch one.

Step 7: Tactic Evaluation: Were the Communications Messages Received?

By determining if the communications messages were received, the success of the tactics can be measured. This can usually be tracked by obtaining metrics on the medium. For example, it is possible to track how many times an article on a website was clicked on or how much engagement a Tweet garnered. If the home institution of the project has a press office, they (as well as partner institutions' press offices) can monitor media inquiries related to the project.

Step 8: Strategy Evaluation: Are the Tactics Working?

To keep the communications strategy current, it should be reviewed approximately every 6 months. Annually, check-in with the representative audience members to see if the communications tactics are addressing the challenges they presented. Be sure to speak to a variety of audience members yearly. With that new information, reprioritize tactics and audiences, and update the strategy.

If the communications efforts have been successful, some of the previous challenges will have been addressed, allowing resources to be focused on new challenges or shift to different tactics in the communications strategy.

PROJECT AND POLICY CHANGES: A PART OF STRATEGY EVALUATION

As the project progresses and reaches the conclusion of certain phases, the communications strategy will need to change to reflect this. Ideally, upcoming changes would be identified during the biannual review of the communications strategy. However, as these changes can happen unexpectedly, they can serve as a makeshift interval at which the communications strategy can be updated.

Once a project or policy change is known, identify the affected audience and review their concerns. Does this change affect those concerns? If so, does it create new challenges or address previous ones? Reaching a significant milestone in the project, like ending sample acquisition, or changes in policy, such as extending publication moratoriums, will likely result in changes to the communications strategy.

For example, once sample acquisition is complete, many of the communications messages to tissue donation sites will no longer be applicable.

By meeting this milestone, messaging and tactics to this group can be relaxed, allowing for resources to be redirected to other communications tactics.

EXAMPLES FROM COMMUNICATIONS STRATEGIES

The following is an example excerpts from an actual communications strategy.

Perception that scientists outside of NIH and TCGA cannot use TCGA data or get involved in TCGA.

Step 1: Who are the audiences and stakeholders?
- Scientists outside TCGA.

Step 2: Challenges: What are their concerns?
- Data from TCGA are only available to people already involved.
- TCGA is a "members only" club.

Step 3: Consideration: Are the audience's concerns valid?
- No, TCGA data are available to the entire cancer research community prior to publication. However, following the Fort Lauderdale principles [1], the TCGA working group has the right to publish the first comprehensive or "global" analysis, provided it is written in a timely fashion.

❏ No, cancer researchers outside of TCGA are encouraged to use the data and are welcome to join TCGA working groups when they can contribute to the analysis (see chapter 7: Pipeline: Data Analysis).

Step 4: Messaging: What should be said in response?
❏ "TCGA data are available to the research community prior to TCGA's marker paper publication."
❏ "Researchers with contributions in mind are invited to join working groups."

Step 5: Tactics: What is the best way to reach the audience?
❏ Clarify the publication and data use policy in the TCGA Publications Guidelines page on the TCGA website.
❏ Increase accessibility of the Publications Guidelines by adding a link to the webpage directly off of the TCGA homepage.
❏ Write a piece on a researcher who is not a funded TCGA investigator and how they became involved in TCGA.
❏ Promote this story in TCGA's quarterly newsletter and Twitter account.

Step 6: Priority: Which audiences and messages are most important?
❏ Amending the publication policy to improve clarity was relatively simple. Likewise, so was making this information available directly off of the homepage. These were two short-term projects.
❏ Writing the profile of the researcher was a longer term project that served two purposes: it provided a real-life example of the communications message and generated new content for the TCGA website and Twitter account.

Step 7: Tactic Evaluation: Were the communications messages received?
❏ When the Publications Guidelines were made accessible directly from the homepage, traffic to the page increased significantly. The peak number of daily page views jumped by 39%.
❏ In the month following the publication of the researcher profile, the webpage was viewed 146 times. From the TCGA homepage, it was the ninth most frequently clicked link.
❏ At the time the researcher profile was published, it was not possible to track visits that came from Twitter or from the TCGA quarterly newsletter. The metrics measurement system was updated a few months later to track that information.

Step 8: Strategy Evaluation: Are the tactics working?
- ❑ Based on engagement with the researcher profile piece and the Publications Guidelines, the tactics seemed to impact the audience. In informal polls and discussions, the audience's perception aligned more closely with TCGA's policy.

The following is an excerpt from a hypothetical communications strategy.

A cancer advocacy group wants multiple myeloma included in the International Cancer Genome Consortium (ICGC).

Step 1: Who are the audiences and stakeholders?
- ❑ Cancer advocacy groups working on multiple myeloma and their constituents who may include patients, researchers, and policy makers.

Step 2: Challenges: What are their concerns?
- ❑ The advocacy group feels that the project is ignoring the significant public health impact of multiple myeloma.

Step 3: Consideration: Are the audience's concerns valid?
- ❑ Yes, the project is not studying multiple myeloma and does not plan to include it. Samples were too difficult to collect and potential collaborators, while engaged, also did not have sufficient samples.

Step 4: Messaging: What should be said in response?
- ❑ "Data generated as part of ICGC's projects will serve as a foundation for research, on which projects to address other cancer types can be built."

Step 5: Tactics: What is the best way to reach the audience?
- ❑ Customize the message to include in publically available information about the project, such as project rationale or "About" section of the homepage.
- ❑ Add this message as a talking point in presentations.
- ❑ Meet with the advocacy group to discuss the project rationale at this time and what might be possible in the future. It may be feasible to include multiple myeloma in a future iteration of the project.
  - ■ Based on the outcomes of this discussion, there may be potential for a blog post or article on the organization's websites.

Step 6: Priority: Which audiences and messages are most important?
- ❑ Consider the size and activity level of the advocacy group.
  - ■ If it is small, meet with the group's leadership and other opinion leaders in the field to convey the message about the project rationale.
  - ■ If it is large and visible, then apply all tactics and consider involving the project's home institution's advocacy relations office or similar group.

Step 7: Tactic Evaluation: Were the communications messages received?
- ❑ Follow-up with advocates after the discussion to see if their concerns have been addressed.
- ❑ Quantify the number clicks and review the comments if a blog piece posted to their website.

Step 8: Strategy Evaluation: Are the tactics working?
- ❑ Follow-up with the relevant advocacy groups and gather their opinions.
- ❑ Monitor the advocacy group's activities to see if the project is still discussed in their advocacy plan, and, if so, what the tone of that discussion is.

COMMUNICATIONS DEVICES: SYMPOSIA AND VISUAL IDENTITY

Annual Scientific Symposia

An annual scientific symposium can serve several communications and project goals. During an annual meeting, project leaders can provide a high-level overview of the project and outline the progress. This "State of the Project Address" will motivate and energize the collaborators by allowing them to see their place in the overarching schema of the project. This fosters a collaborative spirit among team members. This project summary will also remind the group leaders of the overall mission of the project, ensuring that everyone has the same understanding of the project, which is important when team leaders speak with the media. The presentation should have several talking points or "sound bites" for group leaders to appropriate.

These annual scientific meetings can facilitate achievement of programmatic goals. AWG leaders can provide updates about their project to a larger audience, facilitating idea sharing and collaboration. This meeting also serves as a space for group members on the edges of the collaboration to seek additional opportunities to contribute to the project.

Visual Identity

As a support pillar of the communication strategy, a consistent visual identity is important. A visual identity, sometimes called the logo, is the image representation of the project's brand. Creating a visual identity can be an exciting part of developing a communication strategy, but it's important not to get carried away. Visual identities can be ruined with cluttered and overwrought symbolism.

For example, TCGA's visual identity might have included an image that combined the four colors used in DNA sequencing (red, green, blue, and black) for each different word and then merged with the idea of DNA sequence (either Sanger sequencing, a chromatograph, or a karyotype) with a map. As this description shows, visual identities can crash from being overloaded with motifs and metaphors.

Instead, TCGA used a simple text treatment with understated colors and a symbol the combines the idea of a globe, or "Atlas," with DNA, or the "Genome." A text treatment can use a signature color, and sometimes a proprietary font.

THE CANCER GENOME ATLAS

Consistency is important in creating a cohesive brand. The visual identity should be made easily accessible to all members of the project to reduce the likelihood of them imperfectly recreating the visual identity or developing an entirely new one.

■ FIGURE 3.1 A screenshot of the National Cancer Institute's website with the visual identity featured at the top [2].

■ CONCLUSION

A thoughtful communications strategy advances the project's goals by sharing key messages with identified audiences. It unifies messages from project leaders and positions them to take control of what's said about the project. The communications strategy should identify messages prior to choosing a tactic to maximize effectiveness of each tactic. The communications strategy can be developed in a eight-step process: (1) Audiences: Who are the audiences and stakeholders? (2) Challenges: What are their concerns? (3) Consideration: Are the audience's concerns valid? (4) Messaging: What should be said in response? (5) Tactics: What is the best way to reach the audience? (6) Priority: Which audiences and messages are most important? (7) Tactic Evaluation: Were the communications messages received? (8) Strategy Evaluation: Are the tactics working? As the project progresses, it will be necessary to update the communications strategy as milestones are achieved and some policies change. In addition to the tactics identified in Step 5 communications can be advanced through holding scientific symposia and developing a visual identity.

REFERENCES

[1] Sharing data from large-scale biological research projects: a system of tripartite responsibility, <https://www.genome.gov/Pages/Research/WellcomeReport0303.pdf>; [cited September 9, 2015].

[2] Comprehensive cancer information—National Cancer Institute Cancer.gov; [cited September 10, 2015].

Chapter 4

Pipeline: Sample Acquisition

INTRODUCTION

Sample acquisition for a large-scale genomics research project is a lengthy process that varies greatly across studies. The purpose and scale of the study helps define the need for samples. How many samples are required? What types of analytes are needed? What data types will be generated from the samples? What purpose does the data serve? The study design team should set clear objectives and be able to justify the acquisition of new samples for a study. Since public databases are available with vast amounts of mineable data, what is the added value for building a new data set? Below are some reasons to acquire samples and create a data set:

- The study aims cannot be met using currently available raw or processed data.
- Patient tracking is required for the study (eg, cohort or intervention study).
- Creation of a publicly available resource for posterity to use (eg, TCGA, HapMap [1]).

DEFINE THE SAMPLE SET FOR THE PROJECT

Once sample acquisition is justified, the study design team needs to consider the scientific questions they hope to answer through the study. What information will be taken from the samples acquired? What types of genomic characterization are necessary to answer the scientific questions? What level of clinical information or patient history is required to answer the questions? These questions will create some guidelines by which samples can be pursued. In order to answer the scientific questions, the study design must include the type of molecular characterization the samples will undergo (see chapter 5: Pipeline: Data Generation).

The types of samples must also be determined prior to recruitment. If the study involves a disease or other biological affection, will it require matched controls? If so, will they need a germline normal sample or a

Collaborative Genomics Projects: A Comprehensive Guide. DOI: http://dx.doi.org/10.1016/B978-0-12-802143-9.00004-X

tissue normal? Will a matched control patient be required for every affected patient? Perhaps the study requires several samples collected at various time points over the course of an intervention. The data set needs a defined acceptable range of patient descriptors, such as age, race/ethnicity, disease stage, or other demographic/clinical data parameters. A study can choose to focus these criteria based upon known information on the disease, a subpopulation with distinct features of the disease, underrepresentation of a subpopulation in relevant research, or constraints made necessary by the intervention.

When building a sample set the study design team must also strategically determine the minimum sample size to sufficiently power results. Minimizing sample size makes the data set easier to build and reduces the overall amount of risk to patients. This should be done as early as possible in the design process as it sets a requirement around which many study features are based. There are several free online tools available to calculate minimum sample size for study validity [2].

The following sections apply to large-scale genomics research projects involving acquisition of human biospecimen samples for characterization. Note that the term "tissue" may apply to any biospecimen collected. TCGA exclusively collected organ tissue samples, and the terminology used below is consistent with TCGA vocabulary. For further definitions, please refer to Appendix A: Glossary of Terms.

ESTABLISH A CENTRAL BIOSPECIMEN PROCESSING FACILITY

A centralized biospecimen processing facility or biospecimen core resource (BCR) can be used to organize sample processing and serve as a quality control checkpoint for all samples used in the study. A BCR can oversee the acquisition of appropriately consented, standardized, and rigorously collected biospecimens as well as liaise with all contributing biospecimen provider sites. It can also ensure quality transport and preservation of samples. The BCR can perform quality assurance checks on the samples prior to the isolation of analytes to be distributed to data generation center(s). Samples may also be deidentified or anonymized at the BCR with the removal of protected health information in the accompanying clinical data.

The advantages of using a BCR are numerous when receiving samples from multiple biospecimen tissue source sites (TSSs) and distributing to multiple data generation centers. A central processing facility can allow for uniformity in all pre-data generation steps. These steps may include the following:

- DNA/RNA extraction
- Slide preparation
- Analyte processing
- Pathology validation
- Sample integrity check
- Clinical data quality check
- Liaising with sample providers
- Data anonymization
- Other front-end specimen collection processes

Data generation centers may have the capacity to process samples themselves, and in doing so bypass the BCR. However, samples that forgo central processing do not undergo the same rigorous quality checking a BCR provides. They may not be subject to the same unbiased scrutiny and could potentially lack the required patient consents, sample quality criteria, or required clinical/demographic data. This can necessitate a separate auditing process on sample intake practices at the data generation centers which could be a strain on funds and resources.

ESTABLISH SAMPLE QUALIFICATION METRICS

For projects acquiring samples from a biospecimen bank, criteria must be established for sample qualification into the study. Examples of sample requirements to consider include:

- Age of sample
- Preservation method (eg, frozen vs FFPE)
- Clinical parameters (eg, vital status)
- Demographic constraints (eg, age of patient)
- Treatment details (eg, radiation, chemotherapy)

Examples of sample qualification metrics from TCGA included:

- $> 60\%$ tumor nuclei of sample[1]
- Resected sample (no biopsies)
- Primary, untreated tumor
- Paired germline sample available for primary tumor
- Frozen sample after surgery
- Sufficient yield of molecular analytes

---

[1]The percentage of tumor nuclei requirement was relaxed for certain tumor types such as pancreatic cancer due to the low-purity nature of such tumors.

Clinicians and/or disease experts may be employed to aid in qualifying samples. They may require biospecimen slide images, MRI images, pathology reports, etc., to qualify the sample.

If the sample meets minimum qualifications on paper, they should be sent to the BCR. However, an extra set of checks at the BCR should be implemented on the received sample material. If a sample is disqualified, ample records should be kept describing why it was disqualified that is, what fields did not meet the minimum requirements. A decision must be made whether a sample will be partially characterized if it passed minimum qualifications for only one type of molecular analyte required in the study. For example, if a sample passed DNA quality control but fails RNA integrity, should the DNA be characterized?

Another point to consider is whether samples should be collected prospectively, retrospectively, or both? Prospective samples may be collected and stored using the latest methods. However, prospective sample collection can be slow and may not reach minimum sample size requirements for the study. Retrospective samples may already be stored in banks, allowing a sample set to be collected immediately. Disadvantages of retrospective collection include potentially compromised analyte quality based on age/storage methods and lack of consent for desired protocols. If the study has highly specific or unique sample criteria (such as HapMap's requirements on particular geographic communities), a prospective sample set may be advantageous. TCGA allowed both prospective and retrospective cases, which enabled the assembly of a large sample set and kept the door open for clinicians in practice to consent prospective patients for unlimited genomic studies and public data sharing.

All samples collected should have a plan for use, but residual biospecimen can be used later if other data generation platforms prove helpful to the study. The study plan should include provisions for residual biospecimen usage if this door is to remain open. The BCR can be responsible for holding or discarding residual patient samples as agreed upon in the study design.

SAMPLE PROCESSING AND DISTRIBUTION TO DATA GENERATION CENTERS

If the sample does qualify, it needs to be prepared for disbursement to data generation center(s). Sample processing protocols must be established. Steps to consider include:

- How will DNA or protein be isolated?
- How will biospecimen slides be prepared and paraffin blocks be cut?

- What type of tubes will sample be stored in? How will samples be shipped to characterization centers?
- What safeguards will be put in place to avoid sample swaps, contamination, and other human error?
- How will sample aliquots be tracked?

Each plate, tube, slide, or other material distributed needs to be labeled with specific details. Barcode labeling is strongly preferable since it can be scanned, tracked across shipments, and eliminates human transcription error. A human-readable sample identifier (deidentified from the patient) can also be useful for manual tracking and easier data management. Refer to chapter 5 Pipeline: Data Generation for further details on TCGA barcoding.

ESTABLISH CONSENT PROTOCOLS

Patient consent for a multi-site study in which samples may be sent to multiple data generation centers will go through many hands. The consent form must contain language inclusive enough to allow all procedures to be conducted and all potential users of the data to access what is necessary. As the human genome is further understood, the importance of noncoding and other less-understood regions may be realized after the study. If the patient's data is to be used again as knowledge increases, consent language must be inclusive of further studies. If the consent covers "all genomic studies," then it will safely allow for future efforts.

Data sharing for consented patients is also a concern, especially if shared on a public database. In the United States, federal and some state funded studies require submission of data to widely accessible repositories. These databases may be accessible to any institution that abides by the Data Use Agreement for that particular study. If the study requires submission to federal (or other publicly available) databases, the consent form must allow for unlimited sharing of patient data.

In order to gain Institutional Review Board (IRB) approval at US institutions, the study must:

- Minimize risk to subjects and balance them with benefits, if any,
- Obtain informed consent from a party capable of representing the patient, and document appropriately,
- Ensure patient safety when collecting data,
- Protect the welfare of vulnerable populations, and
- Keep selection of subjects equitable, with necessary considerations for vulnerable populations.

Considerations must be made for vulnerable populations, minors, or others unable to consent for themselves if the study involves these individuals. An appropriate guardian must be able to provide consent for those unable to consent themselves. Oral or written assent is to be obtained from minors or adults in compromised situations as applicable; however, assent and/or a guardian witness of assent does not qualify alone as patient consent. Record-keeping of consent documentation must be rigorous; data shared publicly or across multiple centers is complicated and near impossible to recall or destroy if proper consent cannot be proven.

HANDLING INSTITUTIONAL REVIEW BOARDS IN MULTI-CENTER STUDIES

Large-scale genomic studies involving patient data may elect to use a centralized IRB or honor the rulings of each sample provider's institutional IRB. If the study includes patients from multiple sites, the consent form must contain standard language across sites. Each approach has advantages and limitations which must be considered before choosing an approach.

Implications of Using a Centralized IRB

A standing IRB team needs to be established at an institution chosen by the Project Team. Standing meetings would need to be held throughout the sample/site recruitment phase. If international institutions are included in the study, sites need to provide policies and proposals in English for the IRB to review. A centralized IRB ensures sites submit documentation in adherence to US Ethics laws.

Implications of Honoring Individual IRB Approval Rulings

Submission of IRB approval documentation along with material transfer agreement (MTA) and contract development can reduce the amount of paperwork required for onboarding a provider site. The project team or other designated review board can verify sites' IRB documentation as needed. As onboarding paperwork and procedure is the largest deterrent of participation, reduction of these procedures increases the amount of participating sites in a large-scale study. This also requires more complex tracking. Individual sites will need to maintain documentation of compliance with national patient privacy law, principles outlined in the Fort Lauderdale Agreement, and the Common Rule [3].

IDENTIFY POTENTIAL TISSUE SOURCE SITES

Identifying TSSs may be the most difficult step if not already pre-established. If the study requires open solicitation of TSSs, they must undergo rigorous review to ensure they can provide samples of adequate quality and all sample data elements required. Each potential TSS would supply an application with the following details at a minimum:

- Proof of IRB approval
- Proposed full prices for each biospecimen (including a cost breakdown)
- Biospecimen preservation methods
- Experience with sample shipment
- Approximate amount of samples the provider can share
- Material transfer agreement.

For TCGA, these applications were reviewed periodically by an unbiased Source Evaluation Group that reviewed the elements described, negotiated prices, and made protocol recommendations.

ESTABLISH CONTRACTUAL OBLIGATION AND PAYMENT PLANS FOR TISSUE SOURCE SITES

Once TSSs are secured and IRB protocols established, contracts need to be formed for each provider. It must first be determined whether the TSS will be paid for specimens given to the study. Since biospecimen collection, specimen preparation, data organization, quality check, and shipment are time-consuming and involve multiple staff members at the provider site, it is likely the TSS will request payment for their services. Whether or not payment is provided, a contract should be made with the TSS outlining expectations for the logistical and quality requirements of the specimen. If information or biospecimen samples are needed from the TSS over time intervals, the TSS must be incentivized to complete the full course of data collection through the study. If data are collected over a period of time, questions should be attributed to separate data collection forms that are time-point appropriate; this may involve repeat questions when monitoring the status of a health condition, disease onset/development, or effect of intervention.

Example from TCGA: TCGA evolved its contract model over the course of the study's sample collection period. The project aimed to maximize required clinical information gained from the TSS. Information was needed from the TSS upon specimen shipment, upon enrollment into the study, and 1 year following enrollment into the study. Contracts paid a lump sum upon initial sample arrival at the BCR, which facilitated quick and easy sample submission into TCGA. Over time, two major problems were noted: (1) many samples shipped did not meet QC metrics for qualification and (2) follow-up data was difficult to collect and not guaranteed. The contract structure for biospecimen procurement was changed in 2010 and specific to each TSS. Payments for biospecimen were separated into four installments, and based upon receipt of required data and passing of sample quality. Potential disadvantages of this contract model include added paperwork and tracking for TSS personnel, which is unfavorable and intimidating to new collaborators. The inertia to begin providing biospecimens after setting up the contract became even greater. The biospecimen core devoted significantly more effort into follow-up with the TSS to provide requested data, and biospecimen invoicing became a larger task. Overall the contract reform was successful despite the additional administrative burden because money was saved and the sample/clinical data quality improved. After the contract reform, TSSs were now incentivized by money and contractual obligation to provide all required data. Sites now received 25% of contractual payment upon shipment of samples. Other payments were made after the receipt of enrollment and follow-up data. This saved significant amounts of federal money and reduced the waste of biospecimens. Different methods of biospecimen preservation were also studied as they were accepted initially on a trial basis into TCGA.

Fig. 4.1 describes sample shipments per year (occurring 3 months after the data is generated). When biospecimen contracts were reformed in 2010, samples ready for shipment increased. This figure fails to account for the 1179 samples that qualified (75% after 2010) and were not shipped due to acquisition goals being reached and funding constraints for additional tumor samples.

Fig. 4.2 shows the rate of sample acquisition over time per tumor. Only pilot-phase tumor projects were collected before 2009. Many tumors types in the expanded project gained the majority of their samples in late 2010 (lime green) and thereafter.

MANAGEMENT OF CLINICAL DATA COLLECTION

Data collection forms from the TSS will primarily contain patient demographic, diagnostic, and clinical data. Forms need to be developed so that necessary data points can be accurately tracked. Mandatory and optional fields must be designated clearly and should be specific to the TSS. Data quality control must be maintained, as each TSS may use different methods of determining clinical scores or diagnosis.

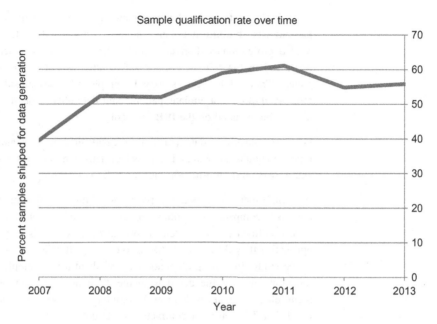

■ **FIGURE 4.1** The improvement of TCGA sample qualification over time is illustrated.

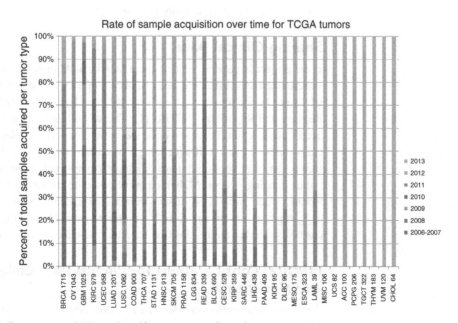

■ **FIGURE 4.2** The improvement of TCGA sample qualification over time is illustrated.

Data collection form development may require the expertise of clinicians, pathologists, bioinformaticians, funders, patient advocates, and other related professionals. Language and/or translation requirements must be addressed in the SOP if clinical information is not in the desired language. Translation services may be employed from outside parties as long as it does not violate patient confidentiality. The translator may have to be included on the IRB protocol.

Original sources of data (such as original slide images and pathology reports) should be obtained if possible. This can serve as a cross-check and may contain information the forms did not encompass.

Assuming that contracts are in place for submission of clinical data from the TSS, compensation policy for required and optional data must be clearly defined and consistent across all centers. Fields may be deemed optional if they do not seem likely to exist at all TSSs but seem relevant to the study. Requiring all fields may result in a low sample size due to potential rarity of the detail from the TSS; however, not requiring too many elements can result in an incomplete and spotty data set. Keep in mind that TSSs are not contractually obligated to provide optional data elements, and pay may not be withheld for refusal to provide optional data. The contract should contain provisions to further the collection of optional data if it proves useful to the analysis. This is usually not apparent before initial genomic analysis is conducted.

Sharing Clinical Data

If the project is publicly funded, there may be additional requirements to standardize clinical data elements (CDEs). For TCGA and other projects, CDEs need to be standardized and searchable in the cancer Data Standards Registry and Repository (caDSR) [4]. This allows for other data users to access public clinical data and build their own data sets according to clinical features of interest. If slide images or other pathological data are collected, they can be shared across multiple pathologists through a secure online imaging system. Allowing remote access to the same slide images would reduce pathologist dissent from minor differences in slides, and save shipment and storage costs. If a protected space to share these images can be established online, this may be more cost-effective and convenient.

Practical Considerations for Clinical Data Collection and Management

A data management system (such as an LIMS) should be employed at the TSS. Access must be regulated such that only relevant users may locate protected information to protect patient confidentiality. Online submission of patient data is ideal for TSS. An online submission apparatus allows for automated first-pass quality checks and prevents omission of required information. A disadvantage for TSS is the time-consuming nature of data submission; inertia may delay the collection of data without repeated persuasion/sufficient incentives. Data fields would ideally be collected in a multiple-choice format in order to conveniently tabulate it downstream. Additional notes may be required from the TSS in free form due to the complicated nature of medical details and treatment techniques. Situations often arise where a patient has a clinical detail not captured on the form, but still qualifies for the study. The ability to report details freely by the TSS can help determine such cases and prevent unnecessary exclusion from the study. The TSS should be trained in the online system used for data submission by the BCR or data collection center. Time must be invested in educating TSS on sample submission, contract requirements, and data collection.

REFERENCES

[1] International HapMap Consortium. The international hapmap project. Nature 2003;426(6968):789–96.
[2] Suresh K, Chandrashekara S. Sample size estimation and power analysis for clinical research studies. J Hum Reprod Sci 2012;5(1):7–13.
[3] Federal policy for the protection of human subjects ('common rule') 2009 [cited May 5, 2015].
[4] caDSR CDE browser [cited May 8, 2015]. Version 4.0.5. Available from: <https://cdebrowser.nci.nih.gov/CDEBrowser/>.

Chapter 5

Pipeline: Data Generation

INTRODUCTION

When the acquisition of new samples is justified for a project, the next step is to generate genomic data from these samples. This is a labor- and resource-intensive activity, therefore the data generation pipeline should be designed very early in the project planning process. Once data has been generated, sufficient storage space and robust data submission pipelines should be implemented. In the following sections, we discuss (1) building a data generation model; (2) establishing a data generation pipeline with data quality measures; (3) proper tracking of data generation and submission. *Note*: Not all projects require submission of data to a public repository; this is determined on an individual project basis.

BUILDING A DATA GENERATION MODEL

It is vital during the early phase of the project to determine (1) what types of genomic data should be generated to accomplish the goals of the project, (2) what technologies and methods should be employed to generate data, and (3) who will be generating the data (centralized vs federated approach). These three aspects will the form the basis of a *data generation model*.

What Data Types Should Be Generated?

The type of data that should be generated for a large-scale genomic study depends on the purpose of the study. For example, in the 1000 Genomes project, one of the aims of the study was to create a catalog of structural variants across many diverse human populations. In order to do this, whole-genome sequencing was performed. One aim of the ENCODE project was to find genome-wide regions of histone modification, thus ChiP-seq (chromatin immunoprecipitation following by high-throughput sequencing) [1] was employed. The "level" of data curation, for example, raw or processed, will also need to be decided upon during project design.

Collaborative Genomics Projects: A Comprehensive Guide. DOI: http://dx.doi.org/10.1016/B978-0-12-802143-9.00005-1

The goal of TCGA was to compile a comprehensive "catalog" of all the molecular changes that occur for over 30 different cancer types. Five main types of molecular data were generated by seven different institutions using validated technologies. The data types were DNA sequence (whole exome and whole genome), RNA sequence, miRNA sequence, DNA methylation, and copy number data, which were generated for each patient sample. For each data type, institutions submitted three general levels of data: "raw," "processed," and "interpreted" data.

Below are descriptions of each level of data.

- *Raw data:* data as it comes from the machine, for example, FASTQ file
- *Processed data:* normalized data, ie, mapping to a reference genome
- *Interpreted:* data interpreted for the presence or absence of a molecular abnormality per sample.

Each data type and data level is summarized in Fig. 5.1.

An in-depth description of each data type, level, and technology used to generate the data is available at the TCGA Data Portal [2].

In general, project stakeholders would access only the "interpreted" data from each data type as that is the level of data that would most likely

Data Type	Level 1	Level 2	Level 3
DNA sequencing (tumor and matched normal control sample)	Sequence reads	Sequence reads aligned to a reference genome	Somatic and germline DNA variants
mRNA sequencing (tumor only)	Sequence reads	Sequence reads aligned to a reference genome	Calculated expression of particular transcript, isoform, exon, and gene
miRNA sequencing (tumor only)	Sequence reads	Sequence reads aligned to a reference genome	Calculated expression of particular miRNA
Copy number data from array (tumor and matched normal control)	Signals per probe or probe set	Normalized copy number data	Copy number data for each region of interest in the genome
DNA Methylation from array (tumor only)	Signals per probe or probe set	Normalized methylation detection values	Methylation detection values mapped to particular regions of the genome

■ FIGURE 5.1 Summary of TCGA data levels and data types.

contain useable information for a project. However, providing the lower level data is crucial as it gives the community of users a chance to reproduce the analysis using their own algorithms and potentially make further discoveries. It is also useful for the development of automated tools and comparison of tools for optimal results.

Summary-level data across a cohort of samples is also important to generate. Examples include significantly mutated genes, mutation rates, etc. Examples of data aggregation used during the analysis phase are available in Chapter 7: Pipeline: Data Analysis.

What Technologies and Methods Should Be Employed to Generate Data?

A discussion of the advantages and disadvantages of available commercial data generation technologies and the best algorithms to analyze data is beyond the scope of this book. Some practical considerations for determining the best technology and methods for data generation and analysis are as follows:

- Direct and indirect costs of procuring, maintaining, and manufacturing for each sequencing or array-based technology
- Direct and indirect costs of data storage and compute
- Added value of implementing an advanced version of a technology when data generation is partially complete
- Data quality metrics of each technology, for example, the number of genes covered by probes of a microarray
- Testing of available algorithms to generate higher level data analysis. Identification of algorithms can be accomplished via a literature search and/or recommendations from a panel of experts in the field
  - Multiple algorithms can be used to detect molecular alterations, and the union or intersection of the findings can be used in analyses. For example, four different algorithms were used to produce TCGA mutation calls from exome sequence data. The concordance between mutation calls was high. This is an effective way of verifying the quality of results of each algorithm and may preclude the need for targeted validation of each detected mutation [3].

Who Will Be Generating the Data?

Another important consideration when building a data generation model is whether the different types of data will be generated by a single center or multiple centers. There are advantages and disadvantages of both methods which are outlined below.

Advantages of single-center data generation:

- The Biospecimen Core Resource (BCR) will not have to generate multiple aliquots of samples for shipment. Aliquoting samples may increase the risk of sample mix-ups (see chapter 4: Pipeline: Sample Acquisition).
- A single tracking system for data generation that is local to the data generating center makes project management more efficient.
- Decrease risk of introducing batch effects [4].
- Promotes uniformity in the data generated.
- Overall decreased cost of data generation (ie economies of scale).

Disadvantages of single-center generation

- Dependence on a single data generation pipeline, which if found to be faulty, will be a blockage in all data production.
- Decreased flexibility in choosing algorithms used for higher level data generation.

Advantages of multi-center data generation

- Presence of alternative pipelines in case one data generation pipeline encounters issues.
- More flexibility in balancing workload for data generation, especially against a deadline.

Disadvantages of multi-center data generation

- More costly as it involves aliquoting samples for distribution.
- Increased risk of sample mix-ups and batch effects, as outlined earlier.
- More administrative burden due to management of multiple institutions.

In TCGA, whole-exome sequencing data was generated at three different centers. This federated approach allowed for data production to continue even if one of the centers was at full capacity. However, the centers did not all follow the same alignment protocol when generating the binary alignment files (.bams) and samples were aligned to different assemblies of the genome. All of the .bams files were eventually realigned to the same assembly; however, this had extra resource costs.

Overall, to determine whether data generation should be centralized or federated, one has to balance between cost, risk, and diversity of algorithms.

ESTABLISHING A DATA GENERATION PIPELINE AND QUALITY CONTROL MEASURES

Once the data generation model has been decided upon, the next step is to build a data generation pipeline. This pipeline generally should have the following four components:

- Sample preparation/nucleic acid extraction
- Sample intake and processing
- Data generation and processing
- Data submission

Quality control measures will need to be developed and assessed at each component of the pipeline. The laboratory protocols for each component will also need to be versioned and kept in an accessible location at each data generation center.

The following are practical considerations for building each component of the pipeline (assuming a single-center data generation model):

- *Sample preparation/nucleic acid extraction.* This can be done at the BCR or the data generation center (see chapter 4: Pipeline: Sample Acquisition). Good Clinical Practices and Good Laboratory Practices [5] must be maintained through the nucleic acid extraction process.
- *Sample intake.* When a sample analyte(s) is shipped to the data generation center(s) from the BCR, a project manager should enter the sample identification numbers in the center's data tracking system (see Proper Tracking of Data Generation section). The sample analyte should be QC'ed (eg, volume, concentration of DNA, etc.). If the sample does not meet these requirements, the project manager should try obtaining more analytes from the BCR (if available) or decide not to generate data for that sample. If an array is used, array protocols should be consistent to avoid sample contamination or swaps.
- *Data generation.* Large-scale projects may involve thousands of samples to be analyzed under time and monetary constraints. For DNA sequencing, multiplexing several samples in one lane has proven cost-effective without compromising data quality in first- and next-generation sequencing [6,7] in the discovery of mutations, copy number variants [8], novel biomarkers, noncoding transcript analysis [9], and validation studies. Pooling sample pairs and other strategies should be employed to minimize variability in each lane.
- *Data submission.* Once the "raw data" has been generated, the data should be stored in a secure local repository at the center. The data should be QC'ed (such as checking the genome coverage of the DNA

sequence reads). Then the data should be formatted to the standards set by the project (see chapter 6: Pipeline: Data Storage and Dissemination). If the project requires "raw data" submission to a centralized data repository, then a secure transfer system should be developed. Best practices for secure data transfer systems can be found in the NIH dbGaP best security practices guide [10].

There should be a team of experts that are responsible for each component of the pipeline. A project manager should communicate between teams and track the progress of the sample through the pipeline. Personnel at each step of the pipeline should have clearly designated duties and responsibilities. Transfer of samples and responsibilities should be communicated to the entire group as each step is completed. This allows for cleaner, faster troubleshooting should an error occur. Clearly designated laboratory space, storage space, and lab equipment can enable better sample tracking.

PROPER TRACKING OF DATA GENERATION

Each center should develop an information tracking system that can track the status of a sample through the data generation pipeline. Most laboratories with high-throughput capacity use a Laboratory Information Management System (LIMS) to track samples. When set up properly, the LIMS system should be able to track:

- Sample identification tag (human or machine-readable barcode) and associated metadata
- Assignment of sample to particular point person and associated analytical workload
- Quality control of the sample or generated sample data
- Equipment utilized to generate data for the sample
- Storage of data generated for the sample and data size
- Inspection, approval, and compilation of sample data for reporting and/or further analysis

■ CONCLUSION

Data generation is perhaps the most varied aspect of a large-scale genomic research project. Technology platforms are constantly being updated. A carefully designed data generation model is essential during the planning phase of the project. An evaluation should be made on the types of data to be generated and the level of curation, the technological approach, and the individual(s) generating the data (single vs multiple institutions).

Afterward, a data generation pipeline should be developed. In each step of the pipeline, quality control checks should be implemented to ensure high-quality data production. Proper tracking and management of data generation and, if required, submission to a central repository is also important in the pipeline design.

REFERENCES

[1] O'Geen H, Echipare L, Farnham PJ. Using ChIP-seq technology to generate high-resolution profiles of histone modifications. Methods Mol Biol 2011;791:265−86.

[2] Data levels and data types, <https://tcga-data.nci.nih.gov/tcga/tcgaDataType.jsp> [cited October 5, 2015].

[3] Davis CF, Ricketts CJ, Wang M, Yang L, Cherniack AD, Shen H, et al. The somatic genomic landscape of chromophobe renal cell carcinoma. Cancer Cell 2014;26(3):319−30.

[4] Leek JT, Scharpf RB, Bravo HC, Simcha D, Langmead B, Johnson WE, et al. Tackling the widespread and critical impact of batch effects in high-throughput data. Nat Rev Genetics 2010;11(10):733−9.

[5] FDA Office of Good Clinical Practice, <http://www.fda.gov/AboutFDA/CentersOffices/OfficeofMedicalProductsandTobacco/OfficeofScienceandHealth Coordination/ucm2018191.htm> [cited October 5, 2015].

[6] Kircher M, Sawyer S, Meyer M. Double indexing overcomes inaccuracies in multiplex sequencing on the Illumina platform. Nucl Acids Res 2012;40(1):e3.

[7] Church GM, Kieffer-Higgins S. Multiplex DNA sequencing. Science 1988;240 (4849):185−8.

[8] Daines B, Wang H, Li Y, Han Y, Gibbs R, Chen R. High-throughput multiplex sequencing to discover copy number variants in *Drosophila*. Genetics 2009;182 (4):935−41.

[9] Laxman B, Morris DS, Yu J, Siddiqui J, Cao J, Mehra R, et al. A first-generation multiplex biomarker analysis of urine for the early detection of prostate cancer. Cancer Res 2008;68(3):645−9.

[10] NIH Security Best Practices for Controlled-Access Data Subject to the NIH Genomic Data Sharing (GDS) Policy, <http://www.ncbi.nlm.nih.gov/projects/gap/pdf/dbgap_2b_security_procedures.pdf> [cited October 5, 2015].

Chapter 6

Pipeline: Data Storage and Dissemination

INTRODUCTION

When a large-scale genomics research project involves many data generation centers and different types of high-volume data, a central data management system needs to be developed. The central data management center serves as the main hub for data upload and download. This is applicable to users within the study and external scientists, as allowed by the project. Its role should include: defining workflows for data management, standardizing data formats, implementing quality control measures, ensuring data security and controlled access, and redistributing data to the project stakeholders and/or the research community. Having a central data management system decreases the risk of data loss and ensures quality and efficiency of data delivery.

CREATION OF CENTRALIZED DATA MANAGEMENT CENTER

A large-scale genomics research project will create a wealth of data that should be centrally managed. A data management or coordination center (DCC) serves as a data storage and distribution center for the project. The DCC ingests data from the data generation centers, stores the data in a structured database, and distributes the data to the data analysis centers and/or to the research community. The role of the DCC should include:

- Work with project stakeholders to define standardized data and metadata formats
- Collect, store, and version data and metadata from various data sources
- Implement quality control measures for submitted data and metadata
- Put appropriate security and access controls in place

Collaborative Genomics Projects: A Comprehensive Guide. DOI: http://dx.doi.org/10.1016/B978-0-12-802143-9.00006-3

■ Redistribute data and metadata tailored to diverse project stakeholders and end users

Depending on the scale of the project, the DCC should have the capability to handle several terabytes of genomic data. For example, the two main databases that store TCGA data: the Data Coordination Center (DCC) and Cancer Genomics Hub (CGHub) ingest 100 TB of data and distribute about 800 TB of data each month to over 8000 unique end users. It is very important that the DCC can meet the technical requirements of project.

DEFINE STANDARD DATA AND METADATA FORMATS

It is vital that data files generated from the project follow a standard format. This optimizes data searching, interpretability, and aggregation from various sources. The format of the data file can be determined by the data type itself. For example, clinical data can be submitted as a simple tab-delimited text file which can be converted into other file formats such as Excel. In the case of genomic sequence files, the FASTQ [1] format is the current universal standard that is used. When the genomic sequences are mapped to a reference genome, the Sequence Alignment Map (SAM) [2] format is generally used. The SAM file is usually converted to a binary form, or Binary Sequence Alignment Map format (BAM) which allows for indexing of sequence alignments. For universally accepted data/metadata formats, the DCC should stay compliant with these standards. Data formatting standards will also change based upon technologies used; it is important for the DCC to keep formats current. The DCC should provide the data submitter with the appropriate tools to submit the data in the defined format.

Higher level analysis files have a less defined format. For example, the TCGA program struggled initially to agree on a standard format for storing mutation data (derived from comparing the tumor and matched normal sample and reference sequences). The Mutation Annotation Format (MAF) (see Appendix D: Mutation Annotation Format (MAF) Specification) was finally conceived as the standard to store this data. It identifies the discovered mutations for each sample and categorizes the type of mutation (eg, SNP, deletion, insertion), origin (somatic or germline), verification status (putative or validated), and any annotations on the mutation. Since the mutation data came from more than three different data generation centers, a standardized format was vital to maximize clarity for users.

If possible, the DCC should define an acceptable type and range of values for each data field for different data types. For example, negative or decimal values should not be allowed in gene expression data for fields that describe the number of transcripts that match to a particular gene region. Free text fields should be minimized as much as possible.

Experimental protocols and analytical tools used to derive the data (ie, metadata) should also be made available in a standardized format for users. This enables users to reproduce the experiment using same input data or their own data. In TCGA, metadata was compiled in a MAGE-TAB (see Appendix E: MAGE-TAB) file that included:

- Investigation Design Format (IDF) file
- Sample and Data Relationship Format (SDRF) file
- Auxiliary description text files of the data

The IDF file includes the name and location of the experimental protocols used to derive the data, any reference data and versions, and contact information for the point person(s) who produced the experiment. The SDRF file related the TCGA sample ID to the data file for that particular sample. However, the DCC did not provide checks on the content of the submitted IDF files and in some cases, important fields lacked sufficient information.

At a minimum, the following metadata should be provided with each data file:

- Experimental protocol and version
- Reference datasets and version
- Descriptions of the fields in each data file and how values were derived
- Contact information (name, email, phone number) for the point person(s) who produced the data

Links to important metadata should be updated (when appropriate) with each subsequent submission.

COLLECT, STORE, AND VERSION DATA AND METADATA

The DCC should have the capability to intake data from many different centers using a standard submission pipeline. The data submission workflow of the TCGA DCC is given in Fig. 6.1.

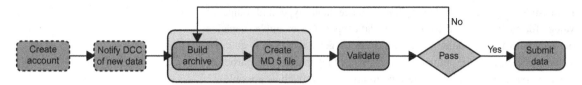

■ **FIGURE 6.1** TCGA DCC submission workflow.

When centers are planning to submit an archive of data, they should alert the DCC via email or internal ticketing system. The centers should create a data file archive based on the specified data format from the DCC. The archive includes all the relevant data and metadata files, and a manifest file describing the contents of the archive. The DCC will then ask the center to upload the file archive via a secure FTP server. Once the upload is complete, the DCC should run various automated QC checks on the contents of the file archive including file format checks, completeness of upload, and compliance with data security policies (see "Implement Quality Control Measures for Submitted Data" section). Should any of the contents of the file archive fail QC check, the DCC should notify the submitting center immediately and work with them to submit corrected files. Only when a data archive is considered error-free, it should be made available to download by various stakeholders.

The DCC should also implement a data versioning system. For long-term projects, data may require multiple modifications and be re-uploaded each time by centers. For example, TCGA clinical data are frequently revised to include additional content, such as follow-up and outcome data. Another important reason for data versioning is accuracy of data set referencing in publications. The data that was used for a publication (see chapter 7: Pipeline: Data Analysis) may not be the most current version; thus, it is important to store historical archives of all data versions for accurate citation and reproducibility efforts.

For any data versioning system, the most recent archive for all data types and levels must be clearly labeled for the users accessing data. There should be no confusion for the users regarding which data version is the most current. All archives should follow a standard archive naming convention (eg, 1.1, 1.2, etc.) and the date and time of upload should be clearly documented.

A data file identifier should also be utilized for easy searching of the file. The identifier should be unique to each file and should be linked to information about the data file such as name, version, submission date, etc.

IMPLEMENT QUALITY CONTROL MEASURES FOR SUBMITTED DATA

An important way to minimize deviations in the data standards as defined by the project is to implement "validators" on the submitted files. These validators should be automated and updated in accordance to changes in file formats, data security polices, etc. The types of checks that validators should be able to do include, but are not limited to:

- File format checks
  - Check that submitted files are named using the correct naming convention.
  - Check that column headers are named consistently between files for the same data type and level.
  - Check that the data values do not deviate from the allowed values and/or range of values.
  - Check that files are not blank, truncated, or have any other content issues.
- Identity checks
  - Check that data files are not duplicates of each other.
  - If data files are submitted for each sample, check that the correct file is submitted for the sample. (This is especially important in a complex project such as TCGA where centers produced data files for over 11,000 samples. Although infrequent, there have been several cases where a center uploaded the incorrect data file for a particular sample. The sample data mismatches were often found during the analysis phase of the project after that file had already been submitted. Tracking the origin of the mismatch and remediation procedures is described further in chapter 8 Quality Control, Auditing, and Reporting.)
  - Check that data file identifier is consistent across revisions.
- Data security checks
  - Check that the data file does not contain protected health information as defined by HIPAA guidelines [3].
  - Check that a file designed as "open-access" does not contain protected information (see "Put in Place Appropriate Security and Access Controls" section).

The above is not an exhaustive list of checks that can be made nor do all projects require such data checks. Validators should be developed in consultation with project stakeholders.

PUT IN PLACE APPROPRIATE SECURITY AND ACCESS CONTROLS

The data management systems developed by the DCC needs to comply with all federal, state, local, and institutional data security policies (for U.S. federal regulations, see Federal Information Management Security Act (FISMA)) [4]. For projects that utilize human subjects research, the systems must maintain compliance with Health Insurance Portability and Accountability Act (HIPAA) Limited Data Set (LDS) specifications. For genomics research projects such as TCGA, there could be individually unique genetic information such as germline information in the form of whole-genome sequences and SNP variant data. NIH policies require that these data be maintained in controlled-access environment. Access to the data in the controlled-access tier requires user authentication under which the user agrees to abide by restrictions on access, use, and redistribution of the data (see TCGA dbGaP User Certification Agreement as an example in Appendix F: Data Use Certification Agreement).

To ensure that certain data that is designated as "restricted access" by the project stakeholders is secure, the DCC should:

- Ensure that this data is stored and maintained in a secure portion of the database
- Establish and implement a user authentication system in which only users with the appropriate credentials are allowed to access this data. Users that violate the data use policies should be immediately removed from access and further access attempts should be closely monitored
- Ensure that this data is downloaded via a secure FTP server in accordance with FISMA standards
- Develop and implement a data management incident protocol in which any improper use of restricted data is reported in a timely manner to various officials and project stakeholders

REDISTRIBUTE DATA AND METADATA TAILORED TO DIVERSE PROJECT STAKEHOLDERS AND END USERS

The DCC should provide an efficient way for users to search and download the data and metadata. This can be as simple as creation of a structured file system for users to download different data archives or as complex as creation of a graphical user interface (GUI) for searching and downloading relevant subsets of data. For larger, complex projects the latter is more appropriate. For GUIs, the following could be elements that are searchable:

- Data type
- Data level
- Version number
- Data submitting center
- Access tier (controlled vs open)
- Sample type
- Sample ID

The retrieval system should package the relevant data files and deliver it to the user as a compressed data archive. For more advanced users, a web Application Programming Interface can be developed. If the project requires a website that is at the top of the application tools, then a content management system (CMS) should be implemented to allow for ordering and tracking of website content and pages.

User documentation needs to be developed and updated as appropriate. Examples of user documentation include:

- User guides on all applications and web interfaces, written for users with diverse levels of biological and bioinformatics expertise
- User guides on data types/file formats and updates when applicable
- Technical documentation such as
  - System architecture and design
  - Databases and database models
  - Software QA plan and test cases
- Data and archive validation specifications

Documentation should be accurate, consistent, up-to-date, comprehensive, and made easily available to the users.

■ CONCLUSION

Having a centralized data management system is advantageous for all stakeholders of the project. Users of the data will be able to retrieve files from one place that have been through a rigorous quality control process. Standard formats allow for easier data aggregation and incorporation into various bioinformatics tools for downstream analysis. This is especially important for data analysis groups in the project. The research community in general benefits from having a central location to download data.

REFERENCES

[1] Cock PJ, Fields CJ, Goto N, Heuer ML, Rice PM. The Sanger FASTQ file format for sequences with quality scores, and the Solexa/Illumina FASTQ variants. Nucleic Acids Res 2010;38(6):1767–71.

[2] Li H, Handsaker B, Wysoker A, Fennell T, Ruan J, Homer N, et al. The sequence Alignment/Map format and SAMtools. Bioinformatics 2009;25(16):2078–9.

[3] Health Information Privacy. <http://www.hhs.gov/ocr/privacy/> [cited October 5, 2015].

[4] Federal Information Security Management Act (FISMA), <http://www.dhs.gov/federal-information-security-management-act-fisma> [cited October 5, 2015].

Chapter 7

Pipeline: Data Analysis

INTRODUCTION

Analysis of a large-scale genomics data set can be daunting unless a plan is in place. As described in chapter 4 Pipeline: Sample Acquisition, developing questions to focus the analysis on is a starting point. A solid background in statistics will also help guide practical efforts in analysis, such as determining whether the results are sufficiently powered and accounting for biases. Tools are constantly emerging for bioinformatics analysis, data visualization, and comparisons between genomic platforms. Having the right set of tools and biological or clinical expertise is crucial to the success of data analysis efforts.

PRECONCEIVED QUESTIONS TO ANSWER

Having a list of scientific questions the data might answer is a starting point in planning data analysis. Many studies are built to test a hypothesis, which requires previously established questions. An exploratory study that does not depend upon outcome data or an expected end result also needs initial questions based upon established knowledge to guide the analysis process. TCGA tumor analysis projects had some questions common to all tumor types, and some individual questions unique to each tumor type. These questions were usually first developed when the clinical data collection forms were created. Questions related to tumor scoring and known risk factors (eg, asbestos exposure and mesothelioma [1]) were included on the forms so that the data would be available for answering these initial questions. Some of the preconceived questions that TCGA Melanoma Analysis Working Group (AWG) explored are described below:

Molecular discovery independent of clinical data:

- Discovery of new driver mutations/translocations/copy number alterations
- Discovery of new driver epigenetic changes
- Identification of cooperating mutational/epigenetic patterns

Collaborative Genomics Projects: A Comprehensive Guide. DOI: http://dx.doi.org/10.1016/B978-0-12-802143-9.00007-5

- Evaluation of mutational signatures (UV exposure, microsatellite instability, etc.)
- Evaluation of new/rare germline SNPs in coding regions that may contribute to risk

Clinical correlation analysis:

- Changes correlated with outcome
  - □ >2 years (or longer with prospective follow-up)
  - □ <2 years
- Changes correlated with organ-specific metastasis
  - □ Brain
  - □ Lung
  - □ Small bowel
  - □ Skin
- Changes correlated with subsequent therapeutic response
  - □ IL2
  - □ Anti-CTLA4
  - □ Chemotherapy/temozolomide

These questions did not end up framing the publication structure. These were starting points for the AWGs to explore; some of these questions yielded poor correlations or statistically insignificant results, while other questions had direct correlations, such as the discovery of improved survival rate of patients with enriched immune gene expression, lymphocyte infiltrate, and high LCK protein expression [2]. Figure 7.1 shows the general workflow of a TCGA tumor analysis project.

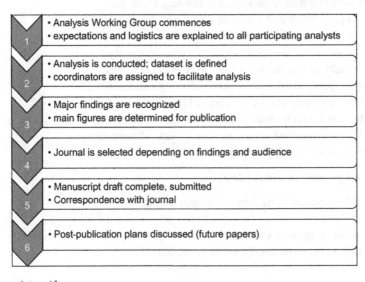

1
- Analysis Working Group commences
- expectations and logistics are explained to all participating analysts

2
- Analysis is conducted; dataset is defined
- coordinators are assigned to facilitate analysis

3
- Major findings are recognized
- main figures are determined for publication

4
- Journal is selected depending on findings and audience

5
- Manuscript draft complete, submitted
- Correspondence with journal

6
- Post-publication plans discussed (future papers)

■ **FIGURE 7.1** TCGA project analysis workflow.

ESTABLISHMENT OF DATA ANALYSIS TEAMS
Analysis Teams and Diversity of Expertise

TCGA organized the analysis of each tumor type into AWGs. These AWGs met every two weeks (or weekly as needed) by teleconference and had a dedicated wiki page where call-related logistical information, agendas, presentation slides, related metadata, and other items were shared. TCGA's AWGs were composed of experts at least 10 different institutions across the country and international participants across several time zones. Teleconferencing and sharing of meeting information in a medium other than email was necessary to consistently share information with all contributors. Each AWG had two leaders: an internal chair from a funded TCGA center and an external chair that was a disease expert on the specific tumor type (usually a tissue provider for TCGA). Other participants included:

- Representatives of each genomic characterization platform
- Representative from the biospecimen core resource (BCR)
- Representative from the data coordination center (DCC)
- Disease experts
- Other analysis experts (eg, statisticians, tumor imaging experts)
- A coordinator and/or manager from the Program Office
- Other Program Officers as needed, including the TCGA Directors
- A data coordinator (role filled by a lead analyst)
- An analysis coordinator (role filled by a lead analyst)
- A manuscript coordinator (role filled by a lead analyst)

The analysis team should have a wide range of skill sets. TCGA AWGs consisted of pathologists, oncologists and other disease specialists, statisticians, sequencing and genomics experts, physician-scientists, epigeneticists, and administrative managers. No one team member was an expert in all project material; each person contributed his or her expertise to the larger project while learning from the others.

Managing Contributions

TCGA eventually had over 1500 participants across all AWGs, with each contributor in at least one AWG. Participants were able to access all the data and were privy to all analysis findings as they developed. The initial goal of the AWGs was to produce a comprehensive, genome-wide analysis publication for each tumor type. Comparable publications of TCGA data from non-AWG researchers were embargoed to allow the AWG an initial publication. Tracking this large amount of participants and their active contributions to the publications was a massive task. Project Managers from the TCGA Program Office regularly reminded participants of the following expectations:

- They are expected to provide significant intellectual contributions to the TCGA publications; just having interest or wanting to learn about the topic is not enough to warrant participant access to results
- Information shared in the AWG meetings and teleconferences (verbal and written) is privileged information. They cannot disseminate the information outside of the group (eg, no data scooping/idea scooping)
- They must adhere to the TCGA Publication Guidelines (see Appendix C: Publication Guidelines as of July 14, 2015) which outline the permissions for independent publication of TCGA data outside the Network while the program is still running

Eventually, the project charter document (see Appendix G: TCGA Analysis Working Group Charter) was revitalized to outline all the rules and expectations of participants. Participants were required to send an email affirming that they read the charter and agree to abide by the rules. Mailing lists were also cleaned out annually to remove participants who had not contributed to the publication since becoming privy to the shared results. Participants were asked to justify their inclusion on the mailing lists, and were removed if sufficient justification was not provided.

ANALYSIS STRUCTURE AND METHODOLOGY
Study Design

Study design will vary depending upon the goals of the project. TCGA was built specifically for discovery and exploration to analyze hundreds of primary untreated tumors of a defined organ system across several different genomic platforms (DNA sequencing, RNA sequencing, miRNA sequencing, copy number variation, protein expression, etc.). Clustering analysis within and across each platform revealed molecular subclasses of the same tumor type. Other studies are designed to validate or expand upon known findings or fit a model that has been theoretically established. Examples are the genome-wide association studies (GWAS) [3] and studies of disease progression/intervention, with multiple samples collected from the same patient at different time points of disease progression/treatment.

When analyzing data across multiple genomic platforms, it is important to compare the platform-specific results against each other. SNP arrays are a widely used tool for sample identity confirmation [4]. Clustering of data within and across platforms can help lead to discovery of different subclasses defined by driver genes, epigenetic change, pathway alterations, treatment response, etc. TCGA analysis benefitted from the development of "cluster of cluster analysis," which helped inform subtyping across tumors based upon gene alterations [5].

Emerging Analytical Tools

As analysis tools are constantly emerging from academic and commercial research organizations, the project must be flexible to allow the use of available tools. TCGA Genome Data Analysis Centers were funded specifically to develop analysis tools, but non-TCGA-funded tools such as HotNet [6] proved very useful for several TCGA tumor projects.

Outliers/Exceptional Cases

TCGA AWGs undoubtedly came across outlier samples that did not fit the subclassifications defined by clustering. Many are quick to remove these outliers from the data set as it may "pollute" cleanly defined data trends or may inhibit the significance of a discovery. TCGA policy instructed analysts to err on the side of inclusion, particularly if the outliers comprise a sample subset that was included in the study by special exception (eg, altered sample qualification requirements). Several steps were taken to determine the nature of the outlying samples, including:

- Review of outlier samples' clinical data to determine any correlated information that would affect clustering behavior (eg, histological subtype, anatomic location, environment details, race/ethnicity, age, sex)
- Review of pathology slides to determine unique pathological features or confirm histology or diagnosis of tumor
- Review of pathology reports, and perhaps discussion with original pathologist, to learn about diagnosis criteria (such as the use of an older classification system for a retrospectively collected tissue) or to ensure that the tumor should not have been excluded (due to chemotherapy, radiation, etc.)
- Comparisons to findings across other platforms to determine how they classify within another platform

These precautions led to several types of discoveries; some samples were found to have been treated, some were found to be misdiagnosed, and yet some were not able to be fully explained from the data provided. For those samples that were later found be disqualified from the study, an annotation was made at the DCC (see chapter 6: Pipeline: Data Storage and Dissemination).

PRACTICAL CONSIDERATIONS
Falling Into a "Formula"

If a large-scale genomics research project includes multiple subprojects with similar structures, care must be taken to maintain flexibility within the structure of analysis methods and publication goals. Analysis strategies in TCGA publications were initially novel and enthusiastically accepted by journals for publication. *Nature* had a verbal agreement with the TCGA Network allowing them the right of first refusal for the first 10 TCGA manuscripts. Only one out of the first 10 TCGA papers was refused by *Nature*. Several years into the project, following the verbal agreement and the emergence of other similar papers on multiplatform analysis, the novelty of TCGA's approach wore off. The approach to analysis had to evolve further in order to remain competitive among similar papers. One approach, taken by some of the rare tumor projects (fewer than 200 samples), was to focus on the angle of clinical applicability. Clustering analyses on a small sample set (such as the TCGA Uterine Carcinosarcoma project, with only 57 tumor samples) was not sufficiently powered to provide meaningful subclasses. Clinical applications, such as therapeutic targets, for a group of similar tumors, were far more useful to the cancer research community.

Aiming for Journals

When analyzing a large data set for publication, the team must eventually decide on a journal which to submit their manuscript. If this decision is made prematurely, it may bias the direction a discovery-type project will go. Will the target audience be clinicians, basic biologists, or genome scientists/bioinformaticians? How many novel findings are in the manuscript? Is it a methods- or technology-heavy manuscript? These determinations, usually occurring after main findings are identified, will help choose the most appropriate journal for the publication. Verbal (informal) or written agreements with journals may also be explored early on if the program would benefit from a series of publications in the same journal, or an issue dedicated to papers produced from the program. Refer to "A Communications Strategy

Applies Tactics to Share Messages" section in chapter 3: Communications Strategies.

Authorship Models

Publications from large-scale genomics research projects can have several hundred contributing authors who all may deserve credit for their work. TCGA employed several authorship models based upon the size of the project:

- *Consortium authorship*—One single author for the paper, for example, The Cancer Genome Atlas Network or The ENCODE Project Consortium [7]. Individual author contributions (determined by the Principal Investigator(s) of each funded TCGA center) are listed in the supplementary information (or other appropriate location depending on the journal) and all authors are indexed in PubMed
- *Traditional authorship*—Authors listed in order of contribution level, with Principal Investigator(s) as last author(s). This was done for projects designated as rare cancers (significantly less samples in the data set than non-rare tumor projects). Contributors may be listed in acknowledgments or the supplementary information but are not indexed in PubMed
- *Consortium and individual authors (ie, mixed)*—The Consortium is listed as a single author within the traditional author list. If this is done, all persons listed as consortium members are indexed in PubMed (Table 7.1)

Timeliness Versus Scientific Merit

The balance between timely publication (before the embargo period ends) and scientific merit was difficult to maintain in TCGA. While TCGA had an obligation as a federally funded study to release data to the public in a timely manner, the data providers also had the right to first publication of their data. The TCGA publication moratoriums were enacted in order to allow the consortium to publish within a decidedly "reasonable" time limit. Difficult decisions had to be made within the AWGs to balance these aspects, particularly in the following:

- When to define the data freeze (data set for publication); analysis began once data was generated on 100 tumors, but new samples kept arriving as sample collection continued
- The major findings to feature in the manuscript, keeping in mind the target audience

Table 7.1 Advantages and Disadvantages of Authorship Models

Authorship Model	Advantages	Disadvantages
Consortium	■ Reduces competition within the program and between projects ■ Provides equal credit for all contributor types for large projects (sample collectors/providers, data managers, genomic data generators, analysts, etc.)	■ Less individual credit for major authors ■ Less incentive to participate for early-career scientists seeking first-author publications unless properly acknowledged for their contributions in another way
Traditional	■ Allows more credit for the most involved authors ■ Promotes name recognition for authors ■ Leads to smaller core group spearheading analysis and ensures active contribution throughout project	■ May be a long list and create competition for prime author slots ■ May not adequately acknowledge all participants if list is culled ■ Difficult to track authors who move institutions during the course of the project
Mixed	■ Allows greater credit for major contributors while acknowledging consortium efforts ■ All contributors are indexed in PubMed if the consortium is an official author	■ May allow for more authors indexed in PubMed than desired ■ Need to determine and enforce clear participation cutoffs to warrant individual author listing, listing within the consortium, or none. May be a point of contention

- Agreeing on the quality of written text for the manuscript and when assembly of the supplementary information was complete
- Whether to include validation data, or to wait on additional (not required) platform data to arrive and incorporate into the study

■ CONCLUSION

Constructing a strategic plan for analyzing large-scale genomic data sets is vital for producing publications for the project. A centralized web page to share information and metadata streamlines communications across the analysis teams and allows easy access to consortium resources. Defined rules of conduct, written agreement from participants, and a high-level plan for managing authorship allows for transparency within the project.

REFERENCES

[1] Wagner JC, Sleggs CA, Marchand P. Diffuse pleural mesothelioma and asbestos exposure in the North Western Cape province. Br J Ind Med 1960;17:260−71.

[2] Cancer Genome Atlas Network. Electronic address IMO, Cancer Genome Atlas N. Genomic classification of cutaneous melanoma. Cell 2015;161(7):1681−96.

[3] Meyer K, Tier B. "SNP Snappy": a strategy for fast genome-wide association studies fitting a full mixed model. Genetics 2012;190(1):275−7.

[4] Liang-Chu MMY, Yu M, Haverty PM, Koeman J, Ziegle J, Lee M, et al. Human biosample authentication using the high-throughput, cost-effective SNPtraceTM system. PLoS One 2015;10(2):e0116218.

[5] Hoadley KA, Yau C, Wolf DM, Cherniack AD, Tamborero D, Ng S, et al. Multiplatform analysis of 12 cancer types reveals molecular classification within and across tissues of origin. Cell 2014;158(4):929−44.

[6] Vandin F, Upfal E, Raphael BJ. Algorithms for detecting significantly mutated pathways in cancer. J Comput Biol J Comput Mol Cell Biol 2011;18(3):507−22.

[7] The ENCODE Project Consortium. An integrated encyclopedia of DNA elements in the human genome. Nature 2012;489(7414):57−74.

Quality Control, Auditing, and Reporting

INTRODUCTION

In a large-scale genomics research project with many different components (sample acquisition, data generation, data storage and dissemination, etc.), it is imperative that standard quality control measures be implemented at each step. This helps minimize deviations from the project protocols and ensures high-quality inputs and outputs from the project. Every component of the pipeline should be using the same central quality management system. The goals of a quality management system should include:

- Establishment of quality metrics for each component of the project pipeline
- Ensure ethical management of samples, information, and derived data sets
- Provide quality reports to stakeholders to help improve processes

ESTABLISH QUALITY METRICS FOR EACH COMPONENT OF THE PIPELINE

Examples of quality metrics have been mentioned throughout the course of this book. It is better to establish quality criteria *before* the project starts that all stakeholders agree on. If multiple institutions are generating the same type of data for a project, each institution should follow the same set of quality standards. Below are some examples of quality metrics that were used in TCGA components:

Sample acquisition (managed by the Biospecimen Core Resource)

- Check if tissue meets minimal percentage of tumor nuclei allowed, and does not exceed maximum necrotic tissue allowed
- Check if tissue can generate the minimum yield of DNA and RNA for genomic analyses
- Check the preservation method for the tissue
- Check if the patient had prior radiation or chemotherapy

Collaborative Genomics Projects: A Comprehensive Guide. DOI: http://dx.doi.org/10.1016/B978-0-12-802143-9.00008-7

- Check the genotype of the tumor, matched tissue normal, and blood sample to ensure that they are derived from the same patient

Data generation (managed by multiple institutions)

- For sequence data, ensure that data meets minimum depth-of-coverage standards
- Check the genotype of the tumor, matched tissue normal, and blood to ensure that they are derived from the same patient (ie, samples were not swapped during the production process). TCGA verified sample identity using the Affymetrix Genome-Wide Human SNP array 6.0
- Check data against control samples to identify any batch effects

Data storage and dissemination (managed by the Data Coordination Center)

- Check that each data file submitted follows the standard data formats
- Check that multiple data files generated for the same patient are associated with the same patient identifier
- Check that descriptive metadata is submitted for each data file archive
- Check that each file can be downloaded without truncation

Quality metrics should be evaluated often during the course of the project to see if they need to be adjusted. However, adjustments should be made with caution and should require the agreement of all stakeholders.

ENSURE ETHICAL MANAGEMENT OF SAMPLES, INFORMATION, AND DERIVED DATA SETS

A quality management system should ensure that samples and derived data sets are managed in accordance to IRB rules and patient consents. For example, for each sample that is qualified by the BCR, the patient consent forms should be checked again to ensure that the use of the sample in the project is within scope. If not, the patient will need to be reconsented. If the patient is deceased, the IRB will need be recontacted for that patient. When the sample is shipped to the data generation center, the quality management system should check to see that an active MTA is in place between the BCR and the data-generating institutions. The system should also check to see if the data types generated for that patient is allowed by the original consent. For example, if a patient was not consented specifically for the identification of certain germline variants, then that data should not be generated. The quality management system should also ensure that access to potentially identifiable data sets is controlled. For every data file that is generated, the system should assess whether the file can be shared in open-access repositories. If not, ensure that it is submitted or distributed in a controlled-access manner. If a patient withdraws consent for the study, then the system should ensure that all patient derived data sets are redacted from study.

PROVIDE QUALITY REPORTS TO STAKEHOLDERS TO HELP IMPROVE PROCESSES

The quality management system should have the ability to report on quality issues and provide detailed documentation on how they were addressed. For example, if a sample failed a certain step in the sample QC process, then the step number, date of assessment, and any rectifying measures should be noted in the system. A unique identification number should be assigned to the quality management incident. Stakeholders of the project should be able to log quality issues of samples or data and search on any quality incidents within the system. A report of quality issues should be provided to relevant stakeholders on a weekly basis and a committee consisting of point people from each component of the project pipeline should review these reports. This helps identify any patterns in quality management incidents that could signal a need to improve the process. For example, if the failure rate of samples are extremely high for a certain provider site, then new sites should be identified for acquiring the tumors. As illustrated in TCGA, if the failure rate for a tumor type is extremely high relative to other tumor types, then tumor nuclearity requirement could potentially be relaxed to allow acceptance of more samples.

EXAMPLE OF A QUALITY MANAGEMENT ISSUE

The below example illustrates how a quality management system can be effective in managing a quality incident for a cancer study similar to TCGA.

Issue: Upon further inspection, a tumor sample that passed quality control measures at the BCR is determined to be a tissue normal (non-tumor) sample.

Steps:

1. Log the incident into the quality management system and assign a unique identifier to the incident.

2. Ascertain whether the sample has already been shipped to the data generation centers.
   a. If no, the BCR will remove the sample from its inventory and document on the system that it has been removed.
   b. If yes, see next step.
3. Ascertain whether data has already been generated for that sample.
   a. If no, the data-generating centers will remove this sample from their pipelines and document in the system that the sample has been removed.
   b. If yes, see next step.
4. The DCC will search for any data files for this sample in the data repository and redact the data files from public download. The DCC will document in the system that the files have been removed.[1] If a publication contains the misclassified data, the data file will be annotated with the error and a corrigendum will be sent to the publication as appropriate.

In each step, the relevant parties will document that a step has been completed for the quality management incident. The end product is a retrievable log that indicates what steps were taken to remediate the incident and the timestamp of each step.

■ CONCLUSION

As illustrated earlier, a centralized quality management system will greatly reduce the administrative burden of tracking the quality outcome of each component of the project pipeline. It optimizes the flow of communication between components of project. The quality management system should also be enabled to track ethical requirements of the project such as whether activities are within scope of the original study consent. The reporting component of the quality management system is especially important as it gives stakeholders the information they need to improve processes.

---

[1]Sometimes the quality management team will decide that sample data should not be removed from the public repository, rather it should just be annotated for users. This approach is justified if the annotation is provided to the users automatically during data download and clearly describes the issue with the sample.

Chapter 9

Project Closure

INTRODUCTION

The endpoint of a project is usually defined by the accomplishment of preconceived project goals. However, endpoints may result from the decision to prematurely cease operations due to various reasons (budgetary and resource constraints, contractual reasons, substantial ethical conflict, etc.). For a large-scale genomics research project, study closure can occur at different levels—publication, institutional, and programmatic. Formal closure protocols should be followed through. Closure planning must be done several months in advance of funding/resource expiration.

LEVELS OF CLOSURE
Publication Level

For many exploratory large-scale research studies such as TCGA, the endpoint is usually marked by the publication of study findings. TCGA was a large program comprised of individual tumor-specific analysis projects. The endpoint of each tumor project was the publication of findings across all available genomic characterization platforms. Federally funded studies in the United States are required to submit all data within the publication to a long-term public data repository. In TCGA, closure of tumor project activities include submission of final publication to the journal, submission and approval of all related media generated by the Program Office (eg, press releases, articles) prior to release date, accounting of all data used for the publication, and submission of publication data to the appropriate repositories.

Institutional Level

Closure at the institutional level can be a lengthy process. For federally funded studies in the United States, each grantee or contracted institution is obligated to complete the deliverables of its grant or contract by a certain timepoint. The deliverables may include delivery of all qualified

Collaborative Genomics Projects: A Comprehensive Guide. DOI: http://dx.doi.org/10.1016/B978-0-12-802143-9.00009-9

tissue samples for biorepository contracts or generation of DNA sequence data of all samples (that pass QC) for genomic characterization centers. Once the deadline is reached, funding for these activities (eg, support for computational biology staff) will end. As an institution prepares to close out a study, grant, or contract, all deliverables must be checked thoroughly and submitted by the date of closure. Plans for continuity or resource reallocation for labor and materials must be developed and ready to implement upon closure date. The following are examples of closure tasks for a genome sequencing center:

- *Identify final deliverables required for study completion and closure.* These may have changed from the beginning of the study, especially when data generation is funded through grants allowing relevant research methods development. A contract may have specified deliverables but may allow flexibility or amendments. It is important to clarify the language of all deliverables when embarking on closure protocols. For example, if a new genome reference model was published during the course of the study and all of the sequences were aligned to the older reference, should all sequence data be remapped to the newest reference? If not, which subset should be remapped? What if all samples were not sequenced using the same technology? TCGA samples were sequenced on a variety of technology platforms over a full decade ABI SOLiD, Illumina, Sanger, Roche 454, and PacBio. Would all samples need to be resequenced on the same platform for uniformity? This would be a resource-intensive requirement for all TCGA samples, but may be doable for a smaller study or a defined subset of TCGA samples. Constraints such as cost, time, resources, data storage, formatting, and residual sample availability are important factors to consider when defining deliverable requirements for study completion and closure.
- *Ensure closure protocols are standardized across all data providers.* For a large-scale genomic study, several data generation centers may be employed to provide the same types of data. TCGA employed hundreds of tissue providers, three DNA sequencing centers, and two RNA sequencing centers. Closure protocols must be consistent across all providers generating data on the same genomic platform. Two examples from TCGA can illustrate different approaches to standardizing closure protocols:
  - TCGA's Genome Sequencing Centers were long-established human genome sequencing centers that had center-wide protocols and QC metrics in practice and development. Over the course of the project, the centers developed their own software for assessing

sample contamination [1,2] and comparisons to SNP data [3]. Each center had its own established cutoff levels for acceptable sample contamination, and its own protocols for how to handle contaminated samples. The centers convened with their grant management institute (National Human Genome Research Institute) to discuss their practices and came to an agreement on how to handle data that did not pass center-specific contamination metrics. It was decided that a minimum cutoff level for contamination was not required for samples to be submitted as part of the closure deliverable. Samples sequenced earlier in the project that may not have passed internal metrics still needed to be submitted with detailed annotations on quality concerns. This example illustrates a difference in practice across centers that was not foreseen or standardized until the end of the study. A risk associated with determining a cross-center standard such as this is record availability. Full completion of this step would be difficult to reach with accuracy if the centers did not keep detailed records of all contamination detection techniques used, contamination scores, decisions made, and reasons.

❏ TCGA tissue providers were also required to furnish original pathology reports and clinical data at the time of biopsy and one year after biopsy. These were all based upon tiered contracts which included payment upon delivery of passable biospecimens and data. Clinical data submission requirements evolved over the years, and some fields that were originally optional became required data elements. Since payment was provided upon receipt of data (as required by the specific contract), information in fields that were originally optional but became mandatory in later iterations could not be guaranteed. The tissue providers were not obligated to provide the information because it was not stipulated as mandatory in their contract, although the material was mandatory on other contracts. This requirement led to some data fields designated as "required" being incomplete across all samples, which proved frustrating to analysts. Despite this, the tissue providers met contractual obligations and their materials were considered complete.

■ *Establish workflow and timelines for closure.* Once all final deliverables are determined and closure protocols established, constructing a timeline will help organize the process. Gantt Charts, Decision Trees, and other project planning tools will be useful for this effort. Input will be needed from all departments involved in the project within the institution to construct a realistic execution strategy.

■ *Prepare final closure documentation for relevant stakeholders.* End-of-project reporting can become complicated if protocols, protocol versioning, and decisions made during the course of the study are not documented. Ideally short-term solutions and long-term protocol changes will have been documented with dates, details, reasons, and samples affected. For a GSC, an ideal final closure report would include the number of samples received, sequenced, QC'ed, and submitted to a data repository and the protocol version used to characterize the samples.

  ❏ For genomic data generation, the machine name, model, and protocol version must be kept individually for each patient. Reagents used, reagent batches, and software used must be recorded clearly. This process may be especially challenging during analysis software or tool development, as versioning may not be as clean during the development process. Eventually, each edition may be binned into versions, and version details with dates active must be kept clearly. Data generation details of all sequenced data in TCGA were submitted in MAGE-TAB format (see Appendix E: MAGE-TAB), which required version names and details of all elements used in the sequencing process. These details needed to be kept over the 9 years, as methods evolved over several areas:
    ■ Sequence generation protocol
    ■ Primary sequencing technology changed from ABI SOLiD to Illumina HiSeq
    ■ Analysis software changes and developments
    ■ Sequence mapping tools
    ■ Genome reference updates
    ■ Different analyses were used on singular or difficult cases
    ■ Protocols to format data into storage file type

  Submission of MAGE-TAB files to the DCC was a deliverable upon submission of the sequence data. Once all data are generated, the final MAGE-TABs had to be delivered with the final batch of data. If detailed records were not kept on each of the steps above, completion of the MAGE-TAB would be nearly impossible.

■ *Close the study for IRB purposes.* Closure documentation for project stakeholders may be repurposed for IRB closure as appropriate, and vice versa. For IRB closure, the following criteria must be met:
  ❏ All data must be generated
  ❏ All data/deliverable products must be submitted

- All information required must be submitted to the stakeholders
- No further efforts will be allocated to the deliverables
- All follow-up activities with patients with regard to the study must be completed
- Any residual materials (biospecimens, treatment data, images, molecular data, etc.) must be handled in accordance with the prior approved protocol. This may allow for material destruction, return, disposal, or retention
- Any remaining obligations agreed before must be met

Although institutions may retain and use clinical data, molecular data, or other information for further studies, these must be completed with regard to the project at hand for IRB closure.

Program Level

The central management office (or Program Office) for the project must track the completion of the deliverables from each grantee or contractor working on the project. The Program Office should help each funded institution to determine a plan of action for the above tasks and to deliver them on time and within budget. After all the data was generated by the genome characterization centers, the TCGA Program Office conducted a full-scale audit of the samples shipped to each center and compared that with the number of samples that were submitted to the data coordination center (DCC). Missing data (data that did not fail center-specific QC measures and should have been submitted) were identified. The missing sample IDs were sent to the characterization centers for investigation and remediation.

The Program Office must also complete all payments to contracted entities. Long-term storage (if applicable) of program-related data and materials should be secured. A program evaluation plan should be developed to ascertain whether the goals have been achieved and that the benefits of the program outweigh the costs. This will determine if funding for future similar projects is warranted.

BUDGETARY CONSIDERATIONS

As a project matures, resources needed to carry out project goals may change due to unexpected complications or developments in methodology or technology. The funding structure defined at the start of the project must be updated throughout the project to reflect current developments.

Budgetary Vigilance by Program Office

TCGA had over 800 tissue providers (TSS) that had pledged to provide an estimated amount of qualified samples at the creation of the contract. Several tissue providers were able to procure more patient samples than were contracted as they encountered more qualified patients than their original estimates. Since some TSS were potentially unable (or lagging) to provide the amount of tissues they estimated, these additional patient samples were welcomed by the TCGA Program Office. As additional patients were added to the study, the total amount of money obligated to the TSS had to increase. The tiered contract system required extra vigilance in making sure that TCGA had the ability to provide money for the tissues. The TCGA Program Office decided to conduct a financial analysis 12 months prior to the end of the sample acquisition period to ensure that tissue providers can be paid for all the tissues they submitted. In some cases, the Program Office had to ask TSS not to provide more tissues than they were contracted (even if they were available) until the program office could confirm that there was enough remaining money to pay for the tissues.

Budgetary Vigilance by Data Providers

TCGA grantees were obligated to provide genomic data or develop tools for data analysis. As the end of the funding period neared, each center had to reassess the priorities of any remaining activities to make sure that primary obligations were met first with the remaining money. For example, the sequencing center may have spent more money than predicted on data generation because samples had to be resequenced or money was used in troubleshooting or process development. Additional projects may have been given to the sequencing center, or projects increased in size over the course of the program. As the funding period draws to a close, all remaining tasks must be reevaluated to determine priorities. Primary tasks must be completed before additional tasks can use project resources.

■ CONCLUSION

As a project draws to a close, plans should be made for the next phase. Genomics is an exciting and dynamic field where analysis efforts must move fast to keep up with competitive academic institutions around the world. Although it is easy to move onto the next step of one project when it is ready, the program must come first. Tasks related to the program must be completed before resources are allocated to the next steps. It is not practical to expect a grantee or an academic institution to focus solely on the present project without preparing for the future; a formal plan must be in place to complete the current project on time and under budget before fully allocating resources to the next steps. Data submission, tissue submission, or delivery of tools/products should be completed for the project outlined in the grant or contract.

REFERENCES

[1] Cibulskis K, McKenna A, Fennell T, Banks E, DePristo M, Getz G. ContEst: estimating cross-contamination of human samples in next-generation sequencing data. Bioinformatics 2011;27(18):2601−2.
[2] ERIS, <https://www.hgsc.bcm.edu/software/eris>. Baylor College of Medicine Human Genome Sequencing Center [cited October 6, 2015].
[3] Koboldt DC, Ding L, Mardis ER, Wilson RK. Challenges of sequencing human genomes. Brief Bioinform 2010;11(5):484−98.

Conclusion

This guide served to recommend best practices on managing large-scale genomics research projects while describing lessons learned through the TCGA program. The authors believe the principles from this guide can be applied to many different types of genomics research projects both in the United States and internationally. The following guiding principles are generalizable to any large consortium research project.

FLEXIBILITY

With any large research project, no amount of planning and organization can account for every challenging situation that may arise. Flexibility is essential. Below are examples of situations that may arise and require temporary or permanent adjustments to the project:

- Development of more advanced methodologies for generating or analyzing data
- Changes in scale (ie, procurement of more samples than originally planned)
- Unforeseen losses of funding, personnel, and/or leadership
- Inability to reach scientific goals due to lack of novel findings or inadequate scientific merit
- Significant changes in costs, such as adapting a new technology, cost inflation, or decrease in technology costs (eg, Moore's law) [1]
- Losing or gaining a contracted organization (ie, data coordination center)
- Encountering unexpected negative publicity

A project must develop creative ways to address these situations. For example, in TCGA funds were allotted to complete a full cancer-specific characterization project—defined as characterizing 500 tumor-normal pairs of a given cancer type on five genomic platforms and collecting clinical data. For rare tumors, the Steering Committee chose to pursue 50 samples of 10 rare tumors types rather than 500 samples of a single disease. Though the costs for sequencing the tumors appear equivalent, the task of carrying out 10 separate projects, assembling different analysis teams, and managing 10 more projects' data was far more resource-intensive.

Collaborative Genomics Projects: A Comprehensive Guide. DOI: http://dx.doi.org/10.1016/B978-0-12-802143-9.00010-5

TRANSPARENCY

Transparency between various stakeholders of the project is essential to maintaining working relationships. As TCGA was a US federally funded program, certain elements of funding appropriation and progress were publicly accessible via online government records or Freedom of Information Act (FOIA) requests. Project statuses, leadership updates, research highlights, and policy updates were updated regularly on the TCGA website. Transparency between funded entities and the funder is also important. In TCGA regular updates on resource allocation, technology development, and pipeline issues were expected by the Program Office from grantee and contractor institutions. Conversely, the Program Office was expected to provide advanced notice of any funding or policy changes to the grantees and contractors.

COLLABORATION

For a large-scale research project requiring a diversity of expertise, collaboration is key. Relationships need to be built and maintained among the personnel of a project to reach project goals. In TCGA participation in analysis working groups were open to the greater scientific community. Although each team had a core group of analysts funded by TCGA, many unfunded researchers joined the analysis teams and provided much needed biological, clinical, or bioinformatics expertise. By opening up participation to the greater scientific community, TCGA fulfilled the aim of building a community resource. TCGA also contributed data to the International Cancer Genomics Consortium (ICGC). Although this created some competition with regard to publication release, it allowed TCGA to combine efforts and share resources on a global scale.

COMMUNICATION

Stakeholders were kept abreast of developments that affected the program through the entire project's duration. Internal stakeholders for TCGA were in large part the Program Office (funders) and funded institutions (grantees and contractors). Regular communication amongst internal stakeholders in carrying out study activities are essential. External stakeholders involved were the patients, pro-bono data providers and analysts, patient advocates, the external scientific committee, and the genomics community at large. Pro-bono data providers were also involved heavily in project-level discussions. TCGA Program Office held biannual meetings to discuss the progress of TCGA and solicit any guidance from the

External Scientific Committee (ESC). Program level updates were provided to the other external stakeholders via TCGA webpage, newsletters, press releases, and other associated press by the NCI office. (Refer to the chapter 3: Communications Strategies for more details on communications strategy). As TCGA was publicly funded, many activities of interest to external stakeholders were public information, eliminating the need for additional efforts to engage external stakeholders. If a study is funded privately or has multiple funding providers, action should be taken to keep the stakeholders appropriately engaged and informed of program developments.

REFERENCE

[1] DNA sequencing costs. Retrieved from NHGRI website <http://www.genome.gov/sequencingcosts/>; October 27, 2015.

Appendix A: Glossary of Terms

Affymetrix manufacturer of the **Genome-wide Human SNP Array 6.0** which was used to generate copy number variation data for TCGA. Please refer to https://tcga-data.nci.nih.gov/tcga/tcgaPlatformDesign.jsp, for a list of array platforms used in TCGA.

Agilent manufacturer of custom gene expression arrays used for TCGA. Gene expression arrays were later replaced by next generation RNA and miRNA sequencing. Please refer to https://tcga-data.nci.nih.gov/tcga/tcgaPlatformDesign.jsp, for a list of array platforms used in TCGA.

Anonymization a method of data protection by which patient privacy is fully protected. Anonymization involves the removal of personally identifiable information from the data set and the assignment of a new identifier. When data are anonymized, the link to reconnect the data back to the patient is broken and the patient cannot be reidentified. This method of privacy protection is best used when no further data will be collected from the patient at a later time.

Array For a list of array technology platforms used in TCGA, please refer to https://tcga-data.nci.nih.gov/tcga/tcgaPlatformDesign.jsp.

BAM a binary version of a tab-delimited text file containing sequence data (SAM). Please refer to https://www.broadinstitute.org/igv/BAM and https://samtools.github.io/hts-specs/SAMv1.pdf, for file format specifications commonly used.

Biospecimen Core Resource (BCR) TCGA employed two BCRs as central hubs to lead the following front-end operations: collect and verify patient clinical data, deidentify patients from the data, receive biospecimens, perform quality control checks on sample identity and integrity, and aliquot and distribute samples to the GCCs and GSCs. The BCRs also maintained clinical data records for patients, collected extra pathological data as necessary, ran pilot studies on the use of FFPE samples for sequencing, and provided annotations/disqualification notes on any sample enrolled. Please refer to http://www.nationwidechildrens.org/biospecimen-core-resource-for-the-cancer-genome-atlas, for more information.

Cancer Genomics Hub (CGHub) long-term repository for storing and cataloging TCGA primary sequence data (FASTQ and BAM files). This includes whole exome, whole genome, mRNA, and miRNA sequence data from all TCGA sequencing sites. See https://cghub.ucsc.edu/, for more information.

Content Management System (CMS) a tool used to store, organize, and publish web content under a unified system. The TCGA Program Office used the Percussion CMS tool to maintain the program's website through the National Cancer Institute.

Data Use Agreement a contract that outlines allowances of data access and use by the scientific community and is essential to maintaining the privacy of genomic data. Users and their home institutions must agree to abide by the data use terms in order to access controlled-access data. Refer to http://cancergenome.nih.gov/pdfs/Data_Use_Certv082014, for TCGA's Data Use Agreement Certification.

Database of Genotypes and Phenotypes (dbGaP) an online database to archive and access genomic data from studies funded by the NIH. Refer to http://www.ncbi.nlm.nih.gov/gap. TCGA controlled-data access was centrally maintained by dbGaP.

Data Coordination Center (DCC) the organization that collects, manages, verifies, stores, and distributes TCGA data within the TCGA Network and greater scientific community. The TCGA DCC was led by the contract organization SRA International.

Deidentification method of data protection by which patient privacy is fully protected. Deidentification involves the removal of personally identifiable information from the data set and the assignment of a new identifier. When data are deidentified, the information needed to reconnect the patient with the data identifier is kept from the public domain and accessed only by authorized data collection personnel. It is used only if necessary for further data collection, and only if the patient is consented to be recontacted.

ENCyclopedia Of DNA Elements (ENCODE) a large-scale genomic research initiative led by the National Human Genome Research Institute (NHGRI) to sequence and catalog all functional parts of the human genome. For more information, please refer to http://www.genome.gov/encode/.

External Scientific Committee an advisory board for TCGA composed of scientists, clinicians, patient advocates and others in the academic, federal, and private industry community with expertise in large scientific programs, genomics, and cancer research. During the initial years of TCGA, they periodically met to provide guidance, direction, and represent community interests in the TCGA effort.

FASTA a format for storing nucleotide sequence data using the one-letter amino acid codes.

FASTQ a format for storing nucleotide sequence data using nucleotide base codes and quality information.

Formalin-fixed, paraffin-embedded (FFPE) a preservation method for biologic tissue in which fresh tissue is preserved immediately upon resection. Formalin fixation prevents tissue and protein degradation, and paraffin embedding stores the tissue in a medium allowing it to be cut and sectioned as necessary. FFPE samples can be stored long-term in room temperature conditions.

Federal Information Security Modernization Act (FISMA) United States federal legislation designed to maintain cyber security of protected information. This extends to sequence or other patient data managed or stored by the federal government, including NIH. Please refer to http://www.dhs.gov/fisma, for more information.

Fresh frozen a preservation method for biologic tissue in which fresh tissue is frozen in −80°C (or similar) conditions immediately upon resection.

The tissue is stored in similarly freezing conditions until it is ready for analysis or nucleic acid extraction.

Functional validation a study used to prove a gene expression or mutation theory by displaying phenotypic effects of the proposed genetic alteration. Nowadays functional validation is normally required to publish genomic findings in scientific journals; however, TCGA Network publications do not contain functional validation exercises.

Genome Characterizations Centers (GCCs) Institutions funded to generate the following genomic data for TCGA samples: copy number variation, gene expression, miRNA expression, protein expression, and DNA methylation.

Genome Data Analysis Centers (GDACs) Institutions funded to develop computational tools to conduct analyses of TCGA data.

Genome mapping the process of aligning raw genomic sequence data with a current human genome reference assembly. TCGA samples were initially mapped to the Genome Reference Consortium human build 36 (GRCh36/HG18) and its variants, but later data generated was mapped to GRCh37/HG19.

Genome sequencing the process of determining the order of nucleotide bases that make up an organism's entire genome. Genome sequencing platforms used in TCGA include ABI SOLiD, Illumina, IonTorrent, Roche 454, Sanger, Complete Genomics, and Pacific Biosciences. Sequencing was done by the **Genome Sequencing Centers (GSCs)** and the **GCCs.**

HapMap a large-scale genomic research project funded by NHGRI to characterize commonly found genetic variations (haplotypes) in human communities. Please refer to http://hapmap.ncbi.nlm.nih.gov/, for more information.

HIPAA the Health Insurance Portability and Accountability Act of 1996 was enacted to protect patient information and ensure privacy. HIPAA regulations are relevant in the United States to the sharing and privacy regulations of genomic data. Please refer to http://www.hhs.gov/ocr/privacy/, for more information.

International Cancer Genomics Consortium (ICGC) a large-scale global genomics consortium focusing on the genome, epigenome, and transcriptome of 50 different tumor types, of which TCGA was a contributor. Please refer to https://icgc.org/, for more information on the consortium.

Investigation Description Format (IDF) a format for an experiment metadata file that includes protocols used, personnel involved, quality control information, experimental variables, etc. It is one of two major components of a Microarray Gene Expression Tabular (MAGE-TAB) format for array and sequence generated. Please refer to https://wiki.nci.nih.gov/display/TCGA/MAGE-TAB;jsessionid = 1E0EC875C92EA6022990722CDF83FD84, for more information on TCGA's documentation of data generated using the MAGE-TAB format, and specifications on the IDF portion.

Illumina manufacturer of next-generation, high-throughput sequencing technologies used by TCGA GSCs and GCCs to generate whole genome, whole exome, mRNA sequence, and miRNA sequence data. Refer to https://www.illumina.com/, for more information on the technology. Illumina also manufactured the **Infinium** array which was used to generate DNA methylation data. TCGA originally used Infinium 27 k array, but switched in

2011 to Infinium 450 k array. Please refer to https://tcga-data.nci.nih.gov/tcga/tcgaPlatformDesign.jsp, for a list of array platforms used in TCGA.

Institutional Review Board (IRB) an ethics committee at a research institution that approves or denies a study from being conducted at the institution based upon ethical judgment of the protocols, data collection forms, patient consent process, endpoints, and other documentation of the study. A multi-site study can either use a central IRB for all sites or honor individual sites' IRB authority and rulings.

Laboratory Information Management System (LIMS) software used to track, organize, and maintain records of laboratory operations.

Mutation annotation file (MAF) a tab-delimited file that contains each mutation of a specific sample, annotated with location, variant type, verification status, etc. For more information on TCGA MAF file specifications, please visit https://wiki.nci.nih.gov/display/TCGA/Mutation + Annotation + Format + (MAF) + Specification.

Microarray Gene Expression Tabular (MAGE-TAB) a file format used to annotate sequence and gene expression data. The two components of a MAGE-TAB are the investigation description format (IDF) and sample and data relationship format (SDRF). Please refer to https://wiki.nci.nih.gov/display/TCGA/MAGE-TAB;jsessionid = 1E0EC875C92EA6022990722CDF83FD84, for more information on TCGA's documentation of data generated using the MAGE-TAB format.

Material transfer agreement (MTA) a contractual agreement between two institutions that involves the transfer of physical materials used for research. The terms of the agreement may include the time frame, scope of project, stipulations for residual material, storage specifications, further transfer protocols, and other limits placed upon the material and/or relationship.

Methylation the bonding of a lone methyl group to a nucleic acid. This occurs in DNA, RNA, histones, and other proteins. DNA methylation affects gene transcription and is one of the required platforms for all TCGA samples.

Mutation an alteration in a protein's nucleotide sequence that may affect gene expression.

Mutation validation a functional or orthogonal procedure to offer proof of mutation detected in sequence data.

Next-generation sequencing a general term for high-throughput sequencing methods that were developed after the original Sanger sequencing method.

Orthogonal validation the validation of mutation data with the use of a different platform or technology than what was used for the original discovery of the mutation.

Pathology report a document from a clinical data source (hospital, cancer center, etc.) with pathological and diagnostic information, completed by the pathologist who received the tissue for diagnosis. TCGA samples required original pathology reports for enrollment.

Patient assent written or oral agreement by a patient to participating in a research and/or medical procedure. Assent is accepted when the patient is unable to provide informed consent, eg, the patient is underage or unable to fully comprehend the consent process. A legal signatory must provide written consent for the patient (along with the patient's assent) in order for procedures to be allowed.

Patient consent voluntary agreement from a patient to participate in a medical or research procedure, or agreeing as a legal guardian for a patient. The consenting individual is of legal age and has the capability of understanding and making the decision to participate in the medical or research procedure.

Program Office the central office in the National Cancer Institute's Center for Cancer Genomics that managed the front end, back end, and day-to-day operations of the TCGA program.

Project Team a group of project responsible for executing and managing a project. The TCGA project team consisted of members from the TCGA Program Office, NHGRI, and the NCI Center for Biomedical Informatics and Information Technology (CBIIT).

Protected Health Information (PHI) patient clinical data, genomic data, or other patient identifying information (such as name, address details, and absolute dates of birth/death) that can be linked back to the specific patient. This information is deidentified or anonymized before the data becomes available to the public to protect patient privacy.

Redaction for TCGA purposes, redaction involves the removal of a case (patient) from the TCGA data set prior to public release or publication on the data. For more information on TCGA redactions, please refer to https://wiki. nci.nih.gov/display/TCGA/Redaction.

Reverse-Phase Protein Array (RPPA) an antibody-based array used to determine protein activities and interactions. RPPA data was generated on a subset of TCGA cases. Please refer to http://www.mdanderson.org/education-and-research/resources-for-professionals/scientific-resources/core-facilities-and-services/functional-proteomics-rppa-core/rppa-process/index.html, for more information on the RPPA procedure.

Sample enrollment the official acceptance of a sample into the TCGA characterization pipeline. This involves creation of an identifier for sample, entrance into the DCC archives, aliquoting and shipment to the GCCs and GSCs for data generation, and request for additional sample data not already submitted for qualification purposes.

Sample qualification The initial process by which a sample is screened for acceptance into the TCGA characterization pipeline. The sample quality is evaluated at the BCR, diagnosis is confirmed by a pathologist, and inclusionary/exclusionary clinical and pathological data is reviewed for acceptable quality.

Sample and Data Relationship Format (SDRF) A tab-delimited reporting format for an array or sequencing project that includes aliquot-level information about data generated in the project. It is one of two major components of a Microarray Gene Expression Tabular (MAGE-TAB) format for representing primary data generated. Please refer to https://wiki.nci.nih.gov/display/TCGA/MAGE-TAB;jsessionid = 1E0EC875C92EA6022990722CDF83FD84, for more information on TCGA's documentation of data generated using the MAGE-TAB format, and specifications on the SDRF portion.

Significantly mutated gene a gene that is mutated in a significant number of patients in a data set.

Single nucleotide polymorphism (SNP) a variation detected at the single nucleotide level commonly found within a species.

Stakeholder any party that has a vested interest in a program or project. This includes project staff, steering committee, contractors/subcontractors, funding entities, patients, participants, or other members of the general public interested in or affected by the project.

Steering Committee a project committee that makes high-level decisions for the program or project, including priorities, next steps, global problem-solving, and current and future directions.

Tissue as defined by TCGA, the term tissue includes all biospecimens collected from a patient (organ tissue, blood, saliva, skin graft, etc.).

VCF variant calling format. This is the universally accepted format for presenting variant calling information as initially defined by the 1000 Genomes project. Please refer to http://www.1000genomes.org/wiki/analysis/variant%20call%20format/vcf-variant-call-format-version-4, for more information.

Wiki The TCGA Program used a wiki page hosted by Atlassian Confluence to organize tumor analysis project information, data formats and specifications, and other data coordination information.

Withdrawal as defined by TCGA, the removal of a case from the study when the patient withdraws consent or no longer agrees to participate in the study. Withdrawal of a patient necessitates redaction of all patient data from the study and the data cannot be published in any way even though it is deidentified.

Appendix B: TCGA Workflow Diagrams

These detailed workflow diagrams represent the reality of data flow and analysis in three major phases as TCGA evolved through 2015.

Fig. B.1. summarizes the deliverables and interplay between the major phases of a TCGA project. Disease Working Groups formed the initial data collection forms and recruited tissues for enrollment and analysis. Analysis Working Groups published global analysis findings (marker papers) on a subset of the enrolled patients. Data and analysis results for the publication sets were stored at the TCGA DCC for public use. All sequence, clinical, and characterization data collected for TCGA are available for public use at the TCGA DCC or CGHub. These storage repositories maintain both public and protected data sets, which require authorization through dbGaP. Governing entities are in black boxes, grant- or contract- funded entities/domains in grey, and unfunded entities/domains in white boxes. Shaded regions encompassing multiple boxes are used to illustrate project phases.

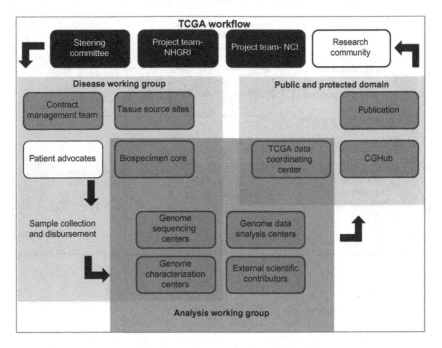

■ FIGURE B.1

Fig. B.2 shows a detailed description of the first phase, the Disease Working Group, and its components and relationships. Tissue provider contracts are formed and managed by the Project Team, Contract Management Team, and Tissue Source Sites (TSS). TSS participants, disease experts, patient advocates, and genomic center representatives work together to form disease-specific data collection forms. With contracts in place, tissues and associated clinical data are collected and managed by the BCR for disbursement to the genomic data generation centers.

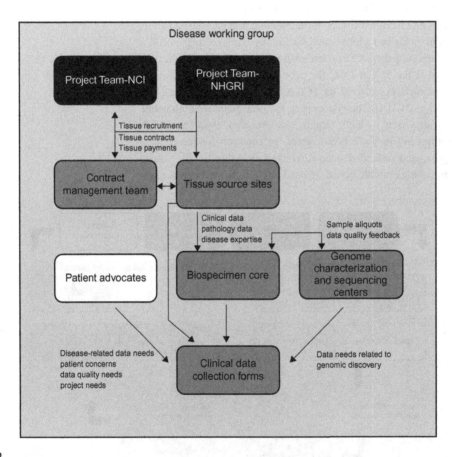

■ FIGURE B.2

Fig. B.3 depicts the Analysis Working Group phase. The two end products of this phase are the submission of data to the DCC (see chapter: Pipeline: Data Generation) and the global analysis publication (see chapter: Pipeline: Data Analysis). Diverse experts from several funded and unfunded entities contribute to these products with scientific guidance, project management, and policy direction from the Project Team and Steering committee. Representatives from the groups are in constant communication with each other, providing data insights, integrated analysis, and publication materials during conference calls and in-person meetings.

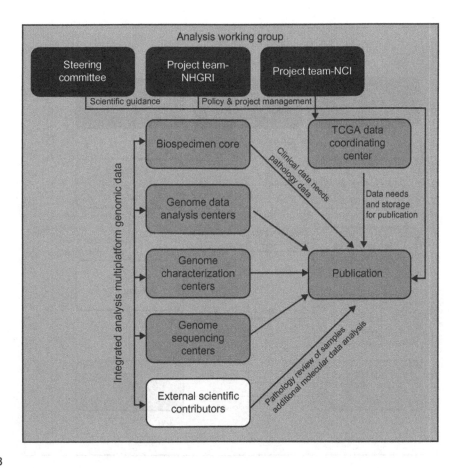

■ FIGURE B.3

Fig. B.4 describes the overarching aim of the TCGA program—a community resource of genomic and clinical data for primary untreated tumor patients. Each type of funded center provides a different platform of data to one of two (formerly three) sources. Sequence and mutation data were initially submitted to the NIH's dbGaP (database of Genotypes and Phenotypes). CGHub was built in 2011 to better house data from multiple sources for such a large project. The Data Coordinating Center provides data quality review, public and protected data curation and distinction, and long-term public access to the data (organized by publication, tumor, or custom data sets).

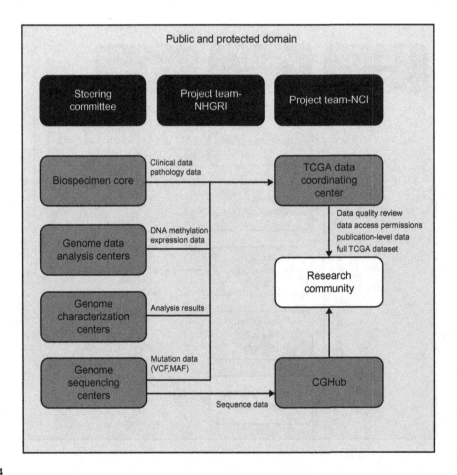

■ FIGURE B.4

As seen in the TCGA example, a simplistic diagram will not depict the full extent of data flow, communication, or deliverable/end products and their relationships. There will always be a need for two-way flow of information and materials at certain phases. Identifying and establishing these relationships will promote communication through the appropriate channels, allowing the most relevant teams to handle questions and complications. When the flow of information is smooth within a large and complicated program, the team may better predict and mitigate issues that arise in data quality, materials expected, accessibility, or scientific integrity.

Appendix C: Publication Guidelines as of July 14, 2015

TCGA is a community resource project, and data are made available rapidly after generation for community research use. To act in accordance with the Fort Lauderdale principles and support the continued prompt public release of large-scale genomic data prior to publication, researchers who plan to prepare reports (publications or presentations) containing descriptions of TCGA data that would be of comparable scope to an initial TCGA comprehensive, global analysis publication, and journal editors or conference organizers who receive such reports are encouraged to coordinate their independent reports with TCGA's publication schedule. *Specifically, comparable scope is defined as global genome-wide analysis of TCGA data from more than one platform OR analysis of data from a single platform across more than one tumor type under moratorium. Publications or presentations involving such analyses are restricted prior to the TCGA publication moratorium release date as described further below.*

TCGA defines a global analysis publication as the first paper authored by The Cancer Genome Atlas Research Network which includes the data from at least 100 cases of a specific tumor type and includes analysis of much of the existing TCGA data on that tumor type at the time. For rare tumor projects a global analysis publication includes data from a majority of the qualified cases and much of the existing data on that tumor type. Specifically, these manuscripts report on the comprehensive, integrated analysis of multiple TCGA data sets available including, but not limited to, copy number variation, gene and miRNA expression, promoter methylation, and DNA sequence/mutation analysis. A global analysis publication is also defined as an analysis of data from a single platform across more than one tumor type under moratorium. Prior to a global analysis publication on a specific tumor type, available data sets should be considered prepublication data subject to the standard principles of scientific etiquette regarding publication of findings using data obtained from other sources.

The TCGA program has established the following policy to clarify freedom of TCGA and non-TCGA users to publish or present on findings using TCGA data. There are no limitations on reports containing analyses using any TCGA data set if the data set meets one of the following three freedom-to-publish criteria:

1. A global analysis publication has been published on that tumor type;
2. 18 months after 100 cases of a given tumor type have shipped from the Biospecimen Core Resource to characterization and sequencing centers or 18 months after final sample shipment for rare tumor projects; or
3. The author receives specific approval from the TCGA Publication Committee in consultation with appropriate disease-specific analysis group(s).

Specifically, the status of each tumor data set is available below. If you have questions, do not hesitate to contact tcga@mail.nih.gov.

Tumor Type	Data Status
Acute myeloid leukemia (AML)	No restrictions; all data available without limitations
Adrenocortical carcinoma (ACC)	No restrictions; all data available without limitations
Breast cancer (BRCA)	No restrictions; all data available without limitations
Cervical cancer (CESC)	No restrictions; all data available without limitations
Chromophobe renal cell carcinoma (KICH)	No restrictions; all data available without limitations
Clear cell kidney carcinoma (KIRC)	No restrictions; all data available without limitations
Colon and rectal adenocarcinoma (COAD, READ)	No restrictions; all data available without limitations
Cutaneous melanoma (SKCM)	No restrictions; all data available without limitations
Diffuse large B-cell lymphoma (DLBC)	No restrictions; all data available without limitations
Esophageal cancer (ESCA)	No restrictions; all data available without limitations
Glioblastoma multiforme (GBM)	No restrictions; all data available without limitations
Head and neck squamous cell carcinoma (HNSC)	No restrictions; all data available without limitations
Liver hepatocellular carcinoma (LIHC)	No restrictions; all data available without limitations

Continued

Continued

Tumor Type	Data Status
Lower grade glioma (LGG)	No restrictions; all data available without limitations
Lung adenocarcinoma (LUAD)	No restrictions; all data available without limitations
Lung squamous cell carcinoma (LUSC)	No restrictions; all data available without limitations
Ovarian serous cystadenocarcinoma (OV)	No restrictions; all data available without limitations
Pancreatic ductal adenocarcinoma (PAAD)	No restrictions; all data available without limitations
Pheochromocytoma and paraganglioma (PCPG)	No restrictions; all data available without limitations
Papillary kidney carcinoma (KIRP)	No restrictions; all data available without limitations
Papillary thyroid carcinoma (THCA)	No restrictions; all data available without limitations
Prostate adenocarcinoma (PRAD)	No restrictions; all data available without limitations
Stomach adenocarcinoma (STAD)	No restrictions; all data available without limitations
Urothelial bladder cancer (BLCA)	No restrictions; all data available without limitations
Uterine carcinosarcoma (UCS)	No restrictions; all data available without limitations
Uterine corpus endometrial carcinoma (UCEC)	No restrictions; all data available without limitations
Projects with limitations until specific dates or until a global analysis is published, whichever comes first. Please contact tcga@mail.nih.gov before publishing.	
Mesothelioma (MESO)	Publication limitations in place until November 14, 2015
Uveal melanoma (UVM)	Publication limitations in place until November 14, 2015
Sarcoma (SARC)	Publication limitations in place until November 28, 2015
Cholangiocarcinoma (CHOL)	Publication limitations in place until November 28, 2015
Testicular germ cell cancer (TGCT)	Publication limitations in place until December 18, 2015
Thymoma (THYM)	Publication limitations in place until December 18, 2015

USE OF TCGA DATA IN PUBLICATIONS AND PRESENTATIONS PRIOR TO INITIAL TCGA GLOBAL ANALYSIS PUBLICATION

Researchers are free, and indeed encouraged, to report results based on TCGA data. These results may include, but would not be limited to, integrating TCGA data with data from other sources, particularly in efforts to study the role of specific genes and genomic changes in the biology of cancer. Researchers also are encouraged to use TCGA data to report on the development of novel methods to analyze genomic data related to cancer and genotype-phenotype relationships in cancer. This may include the application of these methods to portions of the data, for example, specific cancer subtypes, the role of specific genes or gene sets, or particular aspects of tumor biology.

The National Cancer Institute (NCI) and National Human Genome Research Institute (NHGRI) do not consider the deposition of data from TCGA, like those from other large-scale genomic projects, into its own data portal or public databases as the equivalent of publication in a peer-reviewed journal. Therefore, although the data are available to others, the producers consider these data as unpublished and expect that the data will be used in accordance with standard scientific etiquette and practices concerning unpublished data.

Specifically, both TCGA and non-TCGA investigators using TCGA data that have not met a freedom-to-publish criterion per TCGA policy are asked to provide the TCGA Publications Committee and appropriate tumor-specific analysis groups with an abstract summarizing the findings and use of TCGA data. Authors will receive feedback on whether the TCGA Research Network requests that the findings be submitted in coordination with our global analysis publication of a specific tumor type or whether the submission can occur independent of a TCGA publication. This assessment may also apply to submission of manuscripts to journal editors or submission of abstracts to conference organizers. This request can be made via e-mail to tcga@mail.nih.gov.

USE OF TCGA DATA IN PUBLICATIONS AND PRESENTATIONS AFTER INITIAL TCGA GLOBAL ANALYSIS PUBLICATION

There are no restrictions on the use of TCGA data in publications or presentations after the initial TCGA global analysis publication. Once the TCGA Research Network has published tumor-specific analyses, no specific permission is needed for other investigators to publish using these data or future data generated on the same tumor type. Specifically, all data on that tumor type are released from any limitations as soon as the global analysis is published. However, researchers are encouraged to coordinate with the TCGA Research Network via e-mail to tcga@mail.nih.gov when performing multiplatform analysis in these tumor types to explore opportunities for collaboration.

USE OF TCGA DATA FOR RESEARCH PURPOSES OTHER THAN PUBLICATION AND PRESENTATION

There are no restrictions on the use of TCGA data for legitimate research purposes not involving publication or presentation. For example, researchers may use TCGA data in research grant applications or proposals at any time, regardless of whether the initial TCGA global analysis has been published.

TCGA PROGRAM ATTRIBUTION IN PUBLICATIONS AND PRESENTATIONS

TCGA requests that authors who use data from TCGA acknowledge the TCGA Research Network in their work by properly referencing the TCGA data set. Inclusion of the network in the authorship list is not required. However, collaborators seeking to add The Cancer Genome Atlas Research Network to the list of authors should share the manuscript or presentation with the TCGA Steering Committee for approval prior to submission. Manuscripts or presentations may be submitted to tcga@mail.nih.gov for review. Authors are also encouraged to recognize the contribution of the appropriate specimen donors and research groups via the Acknowledgments section in their publication. Similarly, the TCGA program requests that journal editors, reviewers, and conference organizers attempt to ascertain if TCGA is cited and if appropriate acknowledgments are made.

An example of a proper attribution is:

> *The results <published or shown> here are in whole or part based upon data generated by the TCGA Research Network: http://cancergenome.nih.gov/.*

For questions, please contact the TCGA Publications Committee through the Program Office at tcga@mail.nih.gov.

Appendix D: Mutation Annotation Format (MAF) Specification

Mutation Annotation Format (MAF) Specification

The spec has been reverted to the June 26th version (version 20). Additional changes are the removal of the "under construction" banner, changing all text to black, and fixing a typo in the link to the MAF 2.2 specification.

IMPORTANT: MAF files can be submitted to the DCC **only** by following the procedure described here.

Document Information

Specification for Mutation Annotation Format
Version 2.4.1
June 20, 2014

Contents

1 Current version changes
2 About MAF specifications
    2.1 Definition of open access MAF data
    2.2 Somatic MAF vs. Protected MAF
3 MAF file fields
    3.1 Table 1 - File column headers
4 MAF file checks
5 MAF naming convention
6 Previous specification versions

Current version changes

This current revision is **version 2.4.1** of the Mutation Annotation Format (MAF) specification.

The following items in the specification were added or modified in version 2.4.1 from version 2.4:

- Header for MAF file is "#version 2.4.1"
- "Somatic" and "None" are the only acceptable values for "Mutation_Status" for a somatic.MAF (named .somatic.maf). When Mutation_Status is None, Validation_Status must be Invalid.
- Centers need to make sure that Mutations_Status "None" doesn't include germline mutation.
- For a somatic MAF, following rules should be satisfied:
  SOMATIC = (A AND (B OR C OR D)) OR (E AND F)
  A: *Mutation_Status* == "Somatic"
  B: *Validation_Status* == "Valid"
  C: *Verification_Status* == "Verified"
  D: *Variant_Classification* is not {Intron, 5'UTR, 3'UTR, 5'Flank, 3'Flank, IGR}, which implies that *Variant_Classification* can only be \{Frame_Shift_Del, Frame_Shift_Ins, In_Frame_Del, In_Frame_Ins, Missense_Mutation, Nonsense_Mutation, Silent, Splice_Site, Translation_Start_Site, Nonstop_Mutation, RNA, Targeted_Region}.
  E: *Mutations_status* == "*None*"
  F: *Validation_status* == "*Invalid*"
- Extra validation rules: If Validation_Status == Valid or Invalid, then Validation_Method != none (case insensitive).

About MAF specifications

Mutation annotation files should be transferred to the DCC. Those files should be formatted using the mutation annotation format (MAF) that is described below. File naming convention is also below.

Following categories of somatic mutations are reported in MAF files:

- Missense and nonsense
- Splice site, defined as SNP within 2 bp of the splice junction
- Silent mutations

- Indels that overlap the coding region or splice site of a gene or the targeted region of a genetic element of interest.
- Frameshift mutations
- Mutations in regulatory regions

Definition of open access MAF data

A large proportion of MAFs are submitted as discovery data and sites labeled as somatic in these files overlap with known germline variants. In order to minimize germline contamination in putative (unvalidated) somatic calls, certain filtering criteria have been imposed. Based on current policy, open access MAF data should:

- **include**   all validated somatic mutation calls
- **include**   all unvalidated somatic mutation calls that overlap with a coding region or splice site
- **exclude**   all other types of mutation calls (i.e., non-somatic calls (validated or not), unvalidated somatic calls that are not in coding region or splice sites, and dbSNP sites that are not annotated as somatic in dbSNP, COSMIC or OMIM)

Somatic MAF vs. Protected MAF

Centers will submit to the DCC MAF archives that contain Somatic MAF (named **.somatic.maf**) for open access data and an all-inclusive Protected MAF (named **.protected.maf**) that does not filter any data out and represents the original super-set of mutation calls. The files will be formatted using the Mutation Annotation Format (MAF).

The following table lists some of the critical attributes of somatic and protected MAF files and provides a comparison.

Attribute	Somatic MAF	Protected MAF
File naming	Somatic MAFs should be named as ***.somatic.maf** and cannot contain 'germ' or 'protected' in file name.	Protected MAFs should be named as***.protected.maf** and should not contain 'somatic' in the file name.
Mutation category	Somatic MAFs can only contain entries where *Mutation_Status* is "Somatic". If any other value is assigned to the field, the archive will fail. Experimentally validated or unvalidated (see next row) somatic mutations can be included in the file.	There is no such restriction for protected MAF. The file should contain all mutation calls including those from which .somatic.maf is derived.
Filtering criteria	In order to minimize germline contamination, somatic MAFs can contain unvalidated somatic mutations only from coding regions and splice sites, which implies: If *Validation_Status* **is** "Unknown"*Variant_Classification* **cannot** be 3'UTR, 3'Flank, 5'UTR, 5'Flank, IGR, or Intron. *Variant_Classification* can only be \{Frame_Shift_Del, Frame_Shift_Ins, In_Frame_Del, In_Frame_Ins, Missense_Mutation, Nonsense_Mutation, Silent, Splice_Site, Translation_Start_Site, Nonstop_Mutation, RNA, Targeted_Region, De_novo_Start_InFrame, De_novo_Start_OutOfFrame\}. There is no such constraint for experimentally validated (*Validation_Status* is "Valid") somatic mutations. dbSNP sites that are not annotated as somatic in dbSNP, COSMIC or OMIM must be removed from somatic MAFs.	There are no such constraints for mutations in protected MAF.
Access level	These files are deployed as open access data.	These files are deployed as protected data.

MAF file fields

The format of a MAF file is tab-delimited columns. Those columns are described in Table 1 and are required in every MAF file. The order of the columns will be validated by the DCC. Column headers and values **are** case sensitive where specified. Columns may allow null values (i.e._ blank cells) and/or have enumerated values. **The validator looks for a header stating the version of the specification to validate against (e.g. #version 2.4). If not, validation fails.** Any columns that come after the columns described in Table 1 are optional. Optional columns are not validated by the DCC and can be in any order.

Table 1 - File column headers

Index	MAF Column Header	Description of Values	Example
1	Hugo_Symbol	HUGO symbol for the gene (HUGO symbols are *always* in all caps). If no gene exists within 3kb enter "Unknown". Source: http://genenames.org	EGFR
2	Entrez_Gene_Id	Entrez gene ID (an integer). If no gene exists within 3kb enter "0". Source: http://ncbi.nlm.nih.gov/sites/entrez?db=gene	1956
3	Center	Genome sequencing center reporting the variant. If multiple institutions report the same mutation separate list using semicolons. Non-GSC centers will be also supported if center name is an accepted center name.	hgsc.bcm.edu;genome.wustl.edu
4	NCBI_Build	Any TGCA accepted genome identifier. Can be string, integer or a float.-	hg18, hg19, GRCh37, GRCh37-lit e 36.1, 37,
5	Chromosome	Chromosome number without "chr" prefix that contains the gene.	X, Y, M, 1, 2, etc.
6	Start_Position	Lowest numeric position of the reported variant on the genomic reference sequence. Mutation start coordinate (1-based coordinate system).	999
7	End_Position	Highest numeric genomic position of the reported variant on the genomic reference sequence. Mutation end coordinate (inclusive, 1-based coordinate system).	1000
8	Strand	Genomic strand of the reported allele. Variants should always be reported on the positive genomic strand. (Currently, only the positive strand is an accepted value).	+
9	Variant_Classification	Translational effect of variant allele.	Missense_Mutation
10	Variant_Type	Type of mutation. TNP (tri-nucleotide polymorphism) is analogous to DNP but for 3 consecutive nucleotides. ONP (oligo-nucleotide polymorphism) is analogous to TNP but for consecutive runs of 4 or more.	INS
11	Reference_Allele	The plus strand reference allele at this position. Include the sequence deleted for a deletion, or "-" for an insertion.	A
12	Tumor_Seq_Allele1	Primary data genotype. Tumor sequencing (discovery) allele 1. " -" for a deletion represent a variant. "-" for an insertion represents wild-type allele. Novel inserted sequence for insertion should not include flanking reference bases.	C
13	Tumor_Seq_Allele2	Primary data genotype. Tumor sequencing (discovery) allele 2. " -" for a deletion represents a variant. "-" for an insertion represents wild-type allele. Novel inserted sequence for insertion should not include flanking reference bases.	G
14	dbSNP_RS	Latest dbSNP rs ID (dbSNP_ID) or "novel" if there is no dbSNP record. source: http://ncbi.nlm.nih.gov/projects/SNP/	rs12345
15	dbSNP_Val_Status	dbSNP validation status. Semicolon- separated list of validation statuses.	by2Hit2Allele;byCluster
16	Tumor_Sample_Barcode	BCR aliquot barcode for the tumor sample including the two additional fields indicating plate and well position. i.e. TCGA-SiteID-PatientID-SampleID-PortionID-PlateID-CenterID. The full TCGA Aliquot ID.	TCGA-02-0021-01A-01D-0002-04

17	Matched_Norm_Sample_Barcode	BCR aliquot barcode for the matched normal sample including the two additional fields indicating plate and well position. i.e. TCGA-SiteID-PatientID-SampleID-PortionID-PlateID-CenterID. The full TCGA Aliquot ID; e.g. TCGA-02-0021-10A-01D-0002-04 (compare portion ID '10A' normal sample, to '01A' tumor sample).	TCGA-02-0021-10A-01D-0002-04
18	Match_Norm_Seq_Allele1	Primary data. Matched normal sequencing allele 1. "-" for deletions; novel inserted sequence for INS not including flanking reference bases.	T
19	Match_Norm_Seq_Allele2	Primary data. Matched normal sequencing allele 2. "-" for deletions; novel inserted sequence for INS not including flanking reference bases.	ACGT
20	Tumor_Validation_Allele1	Secondary data from orthogonal technology. Tumor genotyping (validation) for allele 1. "-" for deletions; novel inserted sequence for INS not including flanking reference bases.	-
21	Tumor_Validation_Allele2	Secondary data from orthogonal technology. Tumor genotyping (validation) for allele 2. "-" for deletions; novel inserted sequence for INS not including flanking reference bases.	A
22	Match_Norm_Validation_Allele1	Secondary data from orthogonal technology. Matched normal genotyping (validation) for allele 1. "-" for deletions; novel inserted sequence for INS not including flanking reference bases.	C
23	Match_Norm_Validation_Allele2	Secondary data from orthogonal technology. Matched normal genotyping (validation) for allele 2. "-" for deletions; novel inserted sequence for INS not including flanking reference bases.	G
24	Verification_Status	Second pass results from independent attempt using same methods as primary data source. Generally reserved for 3730 Sanger Sequencing.	Verified
25	Validation_Status	Second pass results from orthogonal technology.	Valid

26	Mutation_Status	Updated to reflect validation or verification status and to be in agreement with the VCF VLS field. The values allowed in this field are constrained by the value in the Validation_Status field.	Somatic
27	Sequencing_Phase	TCGA sequencing phase. Phase should change under any circumstance that the targets under consideration change.	Phase_I
28	Sequence_Source	Molecular assay type used to produce the analytes used for sequencing. Allowed values are a subset of the SRA 1.5 library_strategy field values. This subset matches those used at CGHub.	WGS;WXS

29	Validation_Method	The assay platforms used for the validation call. Examples: Sanger_PCR_WGA, Sanger_PCR_gDNA, 454_PCR_WGA, 454_PCR_gDNA; separate multiple entries using semicolons.	Sanger_PCR_WGA;Sanger_PCR_
30	Score	Not in use.	NA
31	BAM_File	Not in use.	NA
32	Sequencer	Instrument used to produce primary data. Separate multiple entries using semicolons.	Illumina GAIIx;SOLID
33	Tumor_Sample_UUID	BCR aliquot UUID for tumor sample	550e8400-e29b-41d4-a716-44665
34	Matched_Norm_Sample_UUID	BCR aliquot UUID for matched normal	567e8487-e29b-32d4-a716-44665

[1] Intergenic Region

[2] 'Consolidated' is used to indicate a site that was initially reported as as variant but subsequently removed from further analysis because it was consolidated into a new variant.   For example, a SNP variant incorporated into a TNP variant.

[3] Used when the discovered variant differs from that of dbSNP

[4] These MAF headers describe the technology that was used to confirm a mutation, whether the same technology ("verification") or a different technology ("validation") is used to prove that a variant is germline or a somatic mutation. [a b]

[5] Explanation of some Validation Status-Mutation Status combinations

Validation Status	Mutation Status	Explanation
Valid	Unknown	a valid variant with unknown somatic status due to lack of data from matched normal tissue.
Invalid	None	validation attempted, tumor and normal are homozygous reference (formerly described as Wildtype)
Inconclusive	Unknown	validation failed, neither the genotype nor its somatic status is certain due to lack of data from matched normal tissue
Inconclusive	None	validation failed, tumor genotype appears to be homozygous reference

> **Important Criteria**
> Index column indicates the order in which the columns are expected. All headers are case sensitive. The Case Sensitive column specifies which values are case sensitive. The Null column indicates which MAF columns are allowed to have null values. The Enumerated column indicates which MAF columns have specified values: an Enumerated value of "No" indicates that there are no specified values for that column; other values indicate the specific values listed allowed; a value of "Set" indicates that the MAF column values come from a specified set of known values (*e.g.* HUGO gene symbols).

MAF file checks

The DCC Archive Validator checks the integrity of a MAF file. Validation will fail if any of the below are not true for a MAF file:

1. Column header text (including case) and order must match specification (Table 1) exactly
2. Values under column headers listed in the specification (Table 1) as not null must have values
3. Values that are specified in Table 1 as Case Sensitive must be.
4. If column headers are listed in the specification as having *enumerated* values (*i.e.* a "Yes" in the "Enumerated" column), then the values under those column must come from the enumerated values listed under "Enumerated".
5. If column headers are listed in the specification as having *set* values (*i.e.* a "Set" in the "Enumerated" column), then the values under

those column must come from the enumerated values of that domain (*e.g.* HUGO gene symbols).
6. All Allele-based columns must contain   - (deletion), or a string composed of the following capitalized letters: A, T, G, C.
7. If Validation_Status   == "Untested"
   then Tumor_Validation_Allele1, Tumor_Validation_Allele2, Match_Norm_Validation_Allele1, Match_Norm_Validation_Allele2 can
   (depending on   Validation_Status).

      a. If Validation_Status   == "Inconclusive"
         then Tumor_Validation_Allele1, Tumor_Validation_Allele2, Match_Norm_Validation_Allele1, Match_Norm_Validation_Allel
         be null (depending on   Validation_Status)
8. If Validation_Status == Valid, then Validated_Tumor_Allele1 and Validated_Tumor_Allele2must be populated (one of A, C, G, T, and -)
      a. If Validation_Status   == "Valid" then Tumor_Validation_Allele1, Tumor_Validation_Allele2, Match_Norm_Validation_Allele1,
         Match_Norm_Validation_Allele2 cannot   be null
      b. If Validation_Status   == "Invalid"
         then Tumor_Validation_Allele1, Tumor_Validation_Allele2, Match_Norm_Validation_Allele1, Match_Norm_Validation_Allel
         nnot be null AND Tumor_Validation_Allele1 == Match_Norm_Validation_Allele1 AND Tumor_Validation_Allelle2 ==
         Match_Norm_Validation_Allele2   (Added as a replacement for 8a as a result of breakdown)
9. Check allele values against Mutation_Status:
   Check allele values against Validation_status:
      a. If Mutation_Status == "Germline" and Validation_Status == "Valid", then Tumor_Validation_Allele1 ==
         Match_Norm_Validation_Allele1 and Tumor_Validation_Allele2 == Match_Norm_Validation_Allele2.
      b. If Mutation_Status == "Somatic" and Validation_Status == "Valid", then Match_Norm_Validation_Allele1 ==
         Match_Norm_Validation_Allele2 == Reference_Allele and (Tumor_Validation_Allele1 or Tumor_Validation_Allele2) !=
         Reference_Allele
      c. If Mutation_Status == "LOH" and Validation_Status=="Valid", then Tumor_Validation_Allele1 == Tumor_Validation_Allele2 and
         Match_Norm_Validation_Allele1 != Match_Norm_Validation_Allele2 and Tumor_Validation_Allele1 ==
         (Match_Norm_Validation_Allele1 or Match_Norm_Validation_Allele2).
10. Check that Start_position <= End_position
11. Check for the Start_position and End_position against Variant_Type:

      a. If Variant_Type == "INS", then (End_position - Start_position + 1 == length (Reference_Allele) or End_position - Start_position
         == 1) and length(Reference_Allele) <= length(Tumor_Seq_Allele1 and Tumor_Seq_Allele2)
      b. If Variant_Type == "DEL", then End_position - Start_position + 1 == length (Reference_Allele), then length(Reference_Allele) >=
         length(Tumor_Seq_Allele1 and Tumor_Seq_Allele2)
      c. If Variant_Type == "SNP", then length(Reference_Allele and Tumor_Seq_Allele1 and Tumor_Seq_Allele2) ==   1 and
         (Reference_Allele and Tumor_Seq_Allele1 and Tumor_Seq_Allele2) != "-"
      d. If Variant_Type == "DNP", then length(Reference_Allele and Tumor_Seq_Allele1 and Tumor_Seq_Allele2) ==   2 and
         (Reference_Allele and Tumor_Seq_Allele1 and Tumor_Seq_Allele2) !contain "-"
      e. If Variant_Type == "TNP", then length(Reference_Allele and Tumor_Seq_Allele1 and Tumor_Seq_Allele2) ==   3 and
         (Reference_Allele and Tumor_Seq_Allele1 and Tumor_Seq_Allele2) !contain "-"
      f. If Variant_Type == "ONP", then length(Reference_Allele) == length(Tumor_Seq_Allele1) == length(Tumor_Seq_Allele2) > 3 and
         (Reference_Allele and Tumor_Seq_Allele1 and Tumor_Seq_Allele2) !contain "-"
12. Validation for UUID-based files:
      a. Column #33 must be Tumor_Sample_UUID containing UUID of the BCR aliquot for tumor sample
      b. Column #34 must be Matched_Norm_Sample_UUID containing UUID of the BCR aliquot for matched normal sample
      c. Metadata represented by Tumor_Sample_Barcode and Matched_Norm_Sample_Barcode should correspond to the UUIDs
         assigned to Tumor_Sample_UUID and Matched_Norm_Sample_UUID respectively
13. If Validation_Status == "Valid" or "Invalid", then Validation_Method != "none" (case insensitive) .

MAF naming convention

In archives uploaded to the DCC, the MAF file name should relate to the containing archive name in the following way:

If the archive has the name

```
<domain>_<disease_abbrev>.<platform>.Level_2.<serial_index>.<revision>.0.tar.gz
```

then a somatic MAF file with the archive should be named according to

```
<domain>_<disease_abbrev>.<platform>.Level_2.<serial_index>[.<optional_tag>].somatic.m
af
```

and a protected MAF with the archive should be named according to

```
<domain>_<disease_abbrev>.<platform>.Level_2.<serial_index>[.<optional_tag>].protected
.maf
```

The `<optional_tag>` may consist of alphanumeric characters, dash, and underscore; no spaces or periods; or it may be left out altogether. The purpose of the optional tag is to impart some brief annotation.

Example

For the archive

```
genome.wustl.edu_OV.IlluminaGA_DNASeq.Level_2.7.6.0.tar.gz
```

the following are examples of valid maf names

```
genome.wustl.edu_OV.IlluminaGA_DNASeq.Level_2.7.somatic.maf
genome.wustl.edu_OV.IlluminaGA_DNASeq.Level_2.7.protected.maf
```

Previous specification versions

- version 2.4
- version 2.3
- version 2.2
- version 2.1
- version 2.0
- version 1.0

Appendix E: MAGE-TAB

Microarray Gene Expression - Tabular format (MAGE-TAB) is a MIAME-compliant, tab-delimited format used to annotate microarray data.

For more information on MAGE-TAB, visit the FGED Society website 🔗 .

Contents

- MAGE Experiments in TCGA

- MAGE-TAB Files

  - Data Files and Data Matrices

- General Notes on Formatting MAGE-TAB Documents

  - MAGE-TAB Archive Validation

  - Data Type Groups that currently use MAGE-TAB

    - Specific Validations

    - Standard Validations

In order to provide a common platform for sharing characterization data within the research community, the Microarray Gene Expression Data (MGED) Society developed the Minimum Information About a Microarray Experiment (MIAME) standard. MIAME describes the data and accompanying metadata that investigators must provide so that the experiment can be reproduced and the results can be interpreted in light of the experimental conditions. MAGE-TAB (MicroArray Gene Expression Tabular) uses simple spreadsheet-based format for representing primary data and associated metadata. MAGE-TAB specification is based on the Microarray and Gene Expression 🔗 Object Model (MAGE-OM 🔗). MAGE-TAB specification 🔗 and related publication 🔗 provide more details on the format.

MAGE Experiments in TCGA

MAGE-based documents usually represent an experiment consisting of many assays. That experiment usually represents a complete study. In the case of TCGA, an **experiment** for a particular center is composed of all the assays of a particular platform for all the samples of a particular tumor type. Since TCGA mandates that data be made available as soon as possible, centers will submit data as soon as possible, so your set of MAGE-TAB documents are required to be updated often for an experiment.

All TCGA characterization data will be modeled using the MAGE-OM and the MAGE-TAB specification will be used to represent the MAGE-OM. One of the goals of modeling and formatting the data is that MAGE-TAB documents can be submitted to external databases (e.g., caArray, ArrayExpress, GEO). Submission of the data is a requirement for its publication and allows querying of the data.

MAGE-TAB Files

To capture experiment details and the relationships between related data files (i.e. data files from different stages of sample data as protocols are continuously applied to it) TCGA uses the MAGE-TAB standard. MAGE-TAB files are tab-delimited text files that model data in the form of columns and rows and is able to capture complex experimental relationships such as an entire study using multiple assays.

MAGE-TAB format uses three different types of files to capture information about an experiment. Click on each file type to learn more about it.

File Type	File Extension	Platform	Description	Example	Mandatory File?
Investigation Description Format (**IDF**)	.idf	mage-tab	Provides general information about the investigation, including its name, a brief description, the investigator's contact details, bibliographic references, and free text descriptions of the protocols used in the investigation.	📄	yes

| Sample and Data Relationship Format (**SDRF**) | .sdrf | mage-tab | Describes the relationships between samples, arrays, data, and other objects used or produced in the investigation, and providing all MIAME information that is not provided elsewhere. In TCGA SDRF files, a row represents an analyzed element (often an aliquot) in its most basic electronic form (i.e. raw data file) and the production of higher-level data files (Level 2 and 3) as protocols (e.g. normalization) are applied to the file and its derivatives. These protocols correspond to those listed in the IDF. | | yes |
| Array Design Format (**ADF**) | .adf | mage-tab | Defines each array type used. An ADF file describes the design of an array, e.g., what sequence is located at each position on an array and what the annotation of this sequence is. An ADF may exist in the MAGE-TAB archive or through the Data Portal on the Platform Design page. | | no |

The following figure depicts the association among different files in a MAGE-TAB archive. The "raw data files" exist in Data Archives.

Data Files and Data Matrices

There may be many different types of data documents including ASCII or binary files (e.g. CEL files), typically in their native formats. A full list of supported formats can be found in the Tab2MAGE data file documentation 🔗.

Preferably data should be provided in a specially defined tab-delimited format termed a "data matrix". MAGE-TAB Data Matrix is a simplified format which allows data columns to be mapped to rows in the SDRF file. The first header line of a Data Matrix file describes this mapping, and the second lists the quantitation types for each column. The first column is used to map the data rows to identifiers from the array design used. MAGE-TAB overview 🔗 and specification 🔗 provide more information and examples.

General Notes on Formatting MAGE-TAB Documents

1. File names of center-specific documents should reflect the archive the file is contained in. Also refer to TCGA Archive Naming Convention.
2. Dates should be formatted as year-month-day (*e.g.* 2007-01-18).
3. Please be careful when editing tab-delimited documents in Microsoft Excel or other spreadsheet applications. Those applications tend to automatically reformat data.
4. Note that the IDF, SDRF, ADF and "data matrix" files should be in plain, tab-delimited text format.
5. The MAGE-TAB specification contains many non-required headers (*e.g.* "Factor Value", "Characteristics", "Protocol Parameters", "Protocol Hardware", "Protocol Software", "Comment", "Normalization Type", "Replicate Type", "Quality Control Types", "Experimental Factor Type", and "Experimental Design"). Please consider adding such values for those headers if the values pertain to your experiment.
6. Please be verbose in your README.txt file and "Experiment Description" MAGE-TAB header.

MAGE-TAB Archive Validation

The DCC uses the ArrayExpress MAGE-TAB scripts in the Tab2MAGE 🔗 software package to create MAGE-ML from MAGE-TAB formatted documents. Therefore, for submitted MAGE-TAB to pass the DCC QC process, it must be successfully processed by the MAGE-TAB scripts. Please keep that in mind before transferring data to the DCC. You may want to run the Tab2MAGE software on your data before transferring it to make sure that MAGE-ML can be created.

Do not reuse example files by adding your IDs. For instance, do not download another center's archive and reuse their IDF or SDRF by adding your own values. Instead, prepare your files according to the design of your center's experiment and the MAGE-TAB specification. It is advised that you run the MAGE-TAB experiment checker 🔗 on your MAGE-TAB documents and also visualize 🔗 the result to see if it makes sense.

Sample MAGE-TAB files
For a real example of IDF and SDRF files, download the MAGE-TAB documents prepared by Memorial Sloan Kettering.

Data Type Groups that currently use MAGE-TAB

Data Type Group (links to specific validations)	Modeled using MAGE-TAB?
aCGH Based Copy Number	Yes

Array Based Expression	Yes
DNA Methylation	Yes
Protein Arrays	Yes
RNASeq and miRNASeq Expression	Yes
SNP-based SNP, Copy Number, LOH	Yes
Low-pass sequencing Based Copy Number	Yes, but in the process of being implemented
GDAC	Yes, but not implemented yet
DNA Sequencing (GSC)	To be specified and implemented

Specific Validations

Validations that are MAGE-TAB specific include

- SDRF Validation
- IDF Validation

 ADF Validations
Note that no ADF validations exist

Standard Validations

MAGE-TAB archives undergo standard validation sets shown in the Standard Archive Validation chart for GCC archives:

- MD5 Validation
- Archive Name Validation
- GCC Experiment Checking
- Manifest Validation
- GCC Experiment Validation
- Data Matrix File Validation
- SDRF Validation
- IDF Validation

Appendix F: Data Use Certification Agreement

The Cancer Genome Atlas (TCGA)

(August 20, 2014 version)

INTRODUCTION AND STATEMENT OF POLICY

The National Institutes of Health (NIH) has established two central data repositories for The Cancer Genome Atlas (TCGA): the Cancer Genomics hub (CGhub) stores lower level sequence data and the TCGA Data Portal stores all other data types and higher level analyses of the sequence data. Systems implemented for the database of Genotypes and Phenotypes (dbGaP) also manage the application, approval, and authentication processes for investigators wishing to access closed access TCGA data. The terms outlined in this Data Use Certification do not apply to open access data. Users should review dbGaP policies for securely storing and sharing human data submitted to NIH under the *Policy for Sharing of Data Obtained in NIH Supported or Conducted Genome-Wide Association Studies (GWAS)*, but users shall note that TCGA data are not GWAS data and have slightly different policies associated with them as outlined in this Data Use Certification. Implicit in the establishment of these data management systems is that scientific progress in genomic research will be greatly enhanced if the data are readily available to all scientific investigators and shared in a manner consistent with the human subjects protocols under which participant's data and samples were donated to TCGA.

Access to human genomic data will be provided to research investigators who, along with their institutions, have certified their agreement with the expectations and terms of access detailed below. It is the intent of NIH, National Cancer Institute (NCI-NIH), and the National Human Genome Institute (NHGRI-NIH) that approved users of TCGA data sets recognize restrictions on data use established by the submitting institution through the Institutional Certification and stated on TCGA's Publication Guidelines page (http://cancergenome.nih.gov/publications/publicationguidelines).

Definitions of terminology used in this document are found in Appendix 1.

The parties to this agreement include: the Principal Investigator (PI) requesting access to the genomic study data set (the "Approved User"), his/her home institution as represented by the Institutional Signing Official designated through the eRA Commons system (the "Requester"), and the relevant NIH Institute or Center (IC). The effective date of this agreement shall be the Project Approval Date, as specified on the Data Access Committee (DAC) approval notification.

TERMS OF ACCESS
Research Use

The requester agrees that if access is approved, (1) the PI named in the Data Access Request (DAR) and (2) those named in the "Senior/Key Person Profile" section of the DAR, including the Information Technology Director or his/her designee, and any trainee, employee, or contractor[1] working on the proposed research project under the direct oversight of these individuals, shall become approved users of the requested data set(s). Research use will occur solely in connection with the approved research project described in the DAR, which includes a one- to two-paragraph description of the research objectives and design. New uses of these data outside those described in the DAR will require submission of a new DAR; modifications to the research project will require submission of an amendment to this application (eg, adding or deleting collaborators from the same institution, adding data sets to an approved project). Access to the requested data set(s) is granted for a period of 1 year as defined below.

Contributing investigators, or their direct collaborators, who provided the data or samples used to generate an NIH genomic data set and who have Institutional Review Board (IRB) approval, if applicable, for broad use of the data are exempt from the limitation on the scope of the research use as defined in the DAR.

---

[1]If contractor services are to be utilized, the principal investigator (PI) requesting the data must provide a brief description of the services that the contractor will perform for the PI (eg, data cleaning services) in the research use statement of the DAR. Additionally, the Key Personnel section of the DAR must include the name of the contractor's employee(s) who will conduct the work. These requirements apply whether the contractor carries out the work at the PI's facility or at the contractor's facility. In addition, the PI is expected to include in any contract agreement requirements to ensure that any of the contractor's employees who have access to the data adhere to the *NIH Policy for Sharing of Data Obtained in NIH Supported or Conducted Genome-Wide Association Studies*, the data use certification agreement, and the dbGaP data.

Requester and Approved User Responsibilities

The requester agrees through the submission of the DAR that the PI named in the DAR has reviewed and understands the principles for responsible research use and data handling of the genomic data sets as defined in the NIH GWAS Data Sharing Policy and as detailed in this Data Use Certification (DUC) agreement. The requester and approved users further acknowledge that they are responsible for ensuring that all uses of the data are consistent with federal, state, and local laws and regulations and any relevant institutional policies. The requester certifies that the approved user is in good standing (ie, no known sanctions) with the institution, relevant funding agencies, and regulatory agencies and is eligible to conduct independent research (ie, is not a postdoctoral fellow, student, or trainee).

Through submission of the DAR, the PI agrees to submit either a project renewal or closeout request prior to the expiration date of the 1-year data access period. The PI also agrees to submit an annual progress update or a final progress report at the 1-year anniversary of the DAR, as described under *Research Use Reporting* below. Failure to submit a renewal or complete the closeout process, including confirmation of data destruction by the Signing Official, may result in termination of all current data access and/or suspension of the approved user and all associated key personnel and collaborators from submitting new DARs for a period to be determined by NIH. Repeated violations or unresponsiveness to NIH requests may result in further measures affecting the requester.

Approved users who may have access to personal identifying information for research participants in the original study at their institution or through their collaborators may be required to have IRB approval. By approving and submitting the attached DAR, the Institutional Signing Official provides assurance that relevant institutional policies and applicable federal, state, and local laws and regulations (if any) have been followed, including IRB approval if required. The Institutional Signing Official also assures through the approval of the DAR that other institutional departments with relevant authorities (eg, those overseeing human subjects research, information technology, or technology transfer) have reviewed the relevant sections of the NIH GWAS Data Sharing Policy and the associated procedures and are in agreement with the principles defined.

NIH anticipates that TCGA data sets will be regularly updated with additional information. Unless otherwise indicated, all statements herein are presumed to be true and applicable to the access and use of all versions of these data sets.

Public Posting of Approved Users' Research Use Statement

PIs agree that if they become approved users, information about themselves, and their approved research use will be posted publicly on the dbGaP website. The information includes the approved user's name and institution, project name, research use statement, and a nontechnical summary of the research use statement.

Security Best Practices Requirements. Note that any scientific collaborators, including contactors, who are not at the same institution as the PI must submit their own DAR.

In addition, citations of publications resulting from the use of NIH genomic data sets may be posted on NIH data repository websites.

Non-Identification

Approved users agree not to use the requested data sets, either alone or in concert with any other information, to identify or contact individual participants from whom data and/or samples were collected. This provision does not apply to research investigators operating with specific IRB approval, pursuant to 45 CFR 46, to contact individuals within data sets or to obtain and use identifying information under an IRB-approved research protocol. All investigators conducting "human subjects research" within the scope of 45 CFR 46 must comply with the requirements contained therein.

Non-Transferability

The requester and approved users agree to retain control of the data and further agree not to distribute data obtained through the DAR to any entity or individual not covered in the submitted DAR. If approved users are provided access to NIH genomic data sets for interinstitutional collaborative research described in the research use statement of the DAR, and all members of the collaboration are also approved users through their home institution(s), data obtained through this DAR may be securely transmitted within the collaborative group.

The requester and approved users acknowledge responsibility for ensuring the review and agreement to the terms within this DUC and the appropriate research use of NIH genomic data by research staff associated with any approved project, subject to applicable laws and regulations. NIH genomic data sets obtained through this DAR, in whole or in part, may not be sold to any individual at any point in time for any purpose.

Approved users agree that if they change institutions during the access period, they will complete the DAR closeout process before moving to their new institution. A new DAR and DUC, in which the new institution agrees to the NIH GWAS data sharing policy, must be approved by the relevant NIH DAC(s) before data access resumes. As part of the closeout process, any versions of the data stored at the prior institution should be destroyed and destruction confirmed in writing by the Signing Official, as described below. However, with advance written notice and approval by the NCI/NHGRI TCGA DAC to transfer responsibility for the approved research project to another approved user from the PI's prior institution, the data do not need to be destroyed.

Data Security and Data Release Reporting

The requester and approved users, including the institutional Information Technology Director or his/her designee, acknowledge NIH's expectation that they have reviewed and agree to handle the requested data set(s) according to the current *dbGaP Security Best Practices*, including its detailed description of requirements for security and encryption. These include, but are not limited to:

- All approved users have completed all required computer security training required by their institution, for example, the http://irtsectraining.nih.gov/, or the equivalent.
- The data will always be physically secured (eg, through camera surveillance, locks on doors/computers, and security guard).
- Servers must not be accessible directly from the internet (eg, they must be behind a firewall or not connected to a larger network) and unnecessary services should be disabled.
- Use of portable media (eg, CD, flash drive, or laptop) is discouraged, but if necessary then they should be encrypted consistent with applicable law.
- Updated antivirus/antispyware software is used.
- Security auditing/intrusion detection software that regularly scans and detects potential data intrusions should be in place.
- Strong password policies for file access are used.
- All copies of the data set are destroyed, as permitted by law and local institutional policies, whenever any of the following occurs:
  - the DUC expires and renewal is not sought;
  - access renewal is not granted;
  - DAC requests destruction of the data set; and
  - continued use of the data would no longer be consistent with the DUC.

In addition, the requester and approved users agree to keep the data secure and confidential at all times and to adhere to information technology practices in all aspects of data management to assure that only authorized individuals can gain access to NIH genomic data sets. This agreement includes the maintenance of appropriate controls over any copies or derivatives of the data obtained through this Data Access Request.

Requesters and approved users agree to notify the NCI/NHGRI TCGA DAC of any unauthorized data sharing, breaches of data security, violations in the presentation and publication embargo period, or inadvertent data releases that may compromise data confidentiality within 24 h of when the incident is identified. As permitted by law, notifications should include any known information regarding the incident and a general description of the activities or process in place to define and remediate the situation fully. Within 3 business days of the NCI/NHGRI TCGA DAC notification, the requester, through the approved user and the Institutional Signing Official, agree to submit to the NCI-NHGRI TCGA Data Access Committee a more detailed written report including the date and nature of the event, actions taken or to be taken to remediate the issue(s), and plans or processes developed to prevent further problems, including specific information on timelines anticipated for action.

All notifications and written reports of data security incidents should be sent to:
tcgadac@mail.nih.gov
NCI/NHGRI TCGA Data Access Committee
Marked: URGENT

The NCI-NIH, NHGRI-NIH, and the NIH, or another entity designated by NIH may, as permitted by law, also investigate any data security incident. Approved users and their associates agree to support such investigations and provide information, within the limits of applicable local, state, and federal laws and regulations. In addition, requesters and approved users agree to work with the NCI/NHGRI TCGA DAC and NIH to assure that plans and procedures that are developed to address identified problems are mutually acceptable and consistent with applicable law.

Intellectual Property

By requesting access to genomic data set(s), the requester and approved users acknowledge the intent of the NIH that anyone authorized for research access through the attached DAR follow the intellectual property (IP) principles in the NIH GWAS Data Sharing Policy as summarized below:

> Achieving maximum public benefit is the ultimate goal of data distribution through the NIH genomic data repositories. The NIH encourages broad use of NIH-supported genotype−phenotype data that is consistent with a responsible approach to management of intellectual property derived from downstream discoveries, as outlined in the NIH's Best Practices for the Licensing of Genomic Inventions and its Research Tools Policy (see http://www.ott.nih.gov/).
> The NIH considers these data as precompetitive and urges approved users to avoid making IP claims derived directly from the genomic data set(s). It is expected that these NIH-provided data and conclusions derived therefrom, will remain freely available, without requirement for licensing. However, the NIH also recognizes the importance of the subsequent development of IP on downstream discoveries, especially in therapeutics, which will be necessary to support full investment in products to benefit the public.

Research Dissemination and Acknowledgment of NIH Genomic Study Data Sets

It is NIH's intent to promote the dissemination of research findings from NIH genomic data set(s) as widely as possible through scientific publication or other appropriate public dissemination mechanisms.

Approved users are strongly encouraged to publish their results in peer-reviewed journals and to present research findings at scientific meetings and are asked to adhere to the TCGA Publication Guidelines and Moratoria (http://cancergenome.nih.gov/publications/publicationguidelines).

Approved users agree to acknowledge the TCGA Research Network in all oral and written presentations, disclosures, and publications resulting from any analyses of the data. A sample acknowledgment statement for The Cancer Genome Atlas data set(s) is as follows:

> *The results published here are in whole or part based upon data generated by The Cancer Genome Atlas managed by the NCI and NHGRI. Information about TCGA can be found at http://cancergenome.nih.gov.*

Research Use Reporting

To assure adherence to NIH policies and procedures for genomic data, approved users agree to provide annual progress updates on how these data have been used, including presentations, publications, and the generation of intellectual property. This information helps NIH evaluate program activities and may be considered by the NIH governance committees as part of NIH's effort to provide ongoing oversight and management of NIH genomic data sharing activities.

Progress updates are provided as part of the annual project renewal or project closeout processes, prior to the expiration of the 1-year data access period. Approved users who are seeking renewal or closeout of a project agree to complete the appropriate online forms and provide specific information such as publications or presentations that resulted from the use of the requested data set(s), a summary of any plans for future research use, any violations of the terms of access described within this DUC and the implemented remediation, and information on any downstream intellectual property generated from the data. Approved users also may include general comments regarding topics such as the effectiveness of the data access process (eg, ease of access and use), appropriateness of data format, challenges in following the policies, and suggestions for improving data access or the program in general.

Note that any inadvertent or inappropriate data release incidents should be reported to the NCI/NHGRI TCGA DAC, according to the agreements and instructions under Term 6.

Non-Endorsement, Indemnification

The requester and approved users acknowledge that although all reasonable efforts have been taken to ensure the accuracy and reliability of TCGA data, the NIH, the NCI/NHGRI TCGA DAC, and contributing investigators do not and cannot warrant the results that may be obtained by using any data included therein. NIH, the NCI/NHGRI TCGA DAC, and all contributors to these data sets disclaim all warranties as to performance or fitness of the data for any particular purpose.

No indemnification for any loss, claim, damage, or liability is intended or provided by any party under this agreement. Each party shall be liable for any loss, claim, damage, or liability that said party incurs as a result of its activities under this agreement, except that NIH, as an agency of the United States, may be liable only to the extent provided under the Federal Tort Claims Act, 28 USC 2671 et seq.

Termination and Violations

This DUC will be in effect for a period of 1 year from the date the data set(s) are made accessible to the approved user ("Approved Access Date"). At the end of the access period, approved users agree to report progress, and renew access or closeout the project. Upon project closure-out, approved users agree to destroy all copies of the requested data set(s), except as required by publication practices or law to retain them.

Copies of NIH genomic data set(s) may not need to be destroyed if, with advance notice and approval by the NCI/NHGRI TCGA DAC, the project has been transferred to another approved user at the same institution. In this case, documentation must be provided that other approved users are using the data set(s) under a DAC-approved DAR.

The requester and approved user acknowledge that the NIH or the NCI/NHGRI TCGA may terminate this agreement and immediately revoke access to all TCGA genomic data sets at any time if the requester is found to be no longer in agreement with the policies, principles, and procedures of the NIH and the NCI/NHGRI TCGA DAC.

By submission of the attached DAR,

- The requester through the Institutional Signing Official attests to the approved users' qualifications for access to and use of NIH genomic data set(s) and agrees to the NIH principles, policies, and procedures for the use of the requested data sets as articulated in this document, including the potential termination of access should any of these terms be violated.
- Requesters and the PI further acknowledge that they have shared this document, and the NIH GWAS Data Sharing Policy, and procedures for access and use of genomic data sets with any approved users, appropriate research staff, and all other key personnel and collaborators identified in the DAR.
- Institutional Signing Officials acknowledge that they have considered the relevant NIH GWAS policies and procedures that they have shared this document and the relevant policies and procedures with appropriate institutional departments, and have assured compliance with local institutional policies related to technology transfer, information technology, privacy, and human subjects research.
- The requestor and PI acknowledge that they reviewed and understand the TCGA-specific publication policies as outlined in section "Research Dissemination and Acknowledgment of NIH Genomic Study Data Sets."
- Institutional Signing Officials acknowledge that their institute is solely responsible for the conduct of all individuals who have access to the data under the DAR, including investigators, contractor staff (both onsite and offsite) and trainees.

APPENDIX 1

Approved User A user approved by the relevant Data Access Committee(s) to access one or more data sets for a specified period of time and only for the purposes outlined in the investigator's approved research use statement. Staff members and trainees under the direct supervision of the approved user are also approved users and must abide by the terms laid out in the Data Use Certificate agreement.

Collaborator An individual who is not under the direct supervision of the principal investigator (PI) (eg, not a member of the PI's laboratory) who assists with the PI's dbGaP research project. Internal collaborators are employees of the requester and work at the same location/campus as the PI. External collaborators are not employees of the requester and/or do not work at the same location as the PI, and consequently must be independently approved to access dbGaP data.

Contributing Investigator An investigator who submitted a genomic data set to an NIH-designated data repository (eg, dbGaP).

Data Access Request (DAR) A request submitted to a Data Access Committee for a specific project specifying the data to which access is sought, the planned research use, and the names of collaborators and the Information Technology Director. The DAR is signed by the investigator requesting the data and her/his institutional signing official. Collaborators and project team members on a request must be from the same institution or organization.

Data Derivative Any data including individual-level data or aggregate genomic data that stems from the original data set deposited (eg, imputed or annotated data) in NIH-designated data repositories (eg, dbGaP). Summary information that is expected to be shared through community publication practices in not included in this term.

Data Use Agreement (DUC) An agreement between the approved user, the requestor, and NIH regarding the terms associated with dbGaP data access and the expectations for use of dbGaP data sets.

dbGaP Approved User Code of Conduct Key principles and practices agreed to by all research investigators requesting access to NIH controlled-access genomic data. The elements within the Code of Conduct reflect the terms of access in the Data Use Certification agreement. Failure to abide by the Code of Conduct may result in revocation of an investigator's access to any and all approved data sets (see https://dbgap.ncbi.nlm.nih.gov/aa/GWAS_Code_of_Conduct.html).

Information Technology (IT) Director Individual with the necessary expertise and authority to vouch for the IT capacities at an academic institution, company, or other research entity and the ability of that institution to comply with NIH data security expectations. The IT Director is to be included as key personnel in the Data Access Request.

Institutional Certification Certification by the Institution that delineates, among other items, the appropriate research uses of the data and the uses that are specifically excluded by the relevant informed consent documents (see http://grants.nih.gov/grants/guide/notice-files/NOT-OD-07-088.html).

Institutional Signing Official   Generally, a senior official at an institution with the authority to sign on behalf of the submitting investigator or an investigator who has submitted a Data Access Request or Project Request to NIH, authorized to enter their institution into a legally binding contract, and who is credentialed through the eRA Commons system.

Progress Update   Information included with the annual Data Access Request (DAR) renewal or closeout summarizing the analysis of dbGaP data sets obtained through the DAR and any publications and presentations derived from the work.

Project Closeout   Closeout of a research project that used controlled-access data from an NIH-designated data repository (eg, dbGaP) and confirmation of data destruction when the research is completed and/or discontinued (see http://www.ncbi.nlm.nih.gov/books/NBK63627/).

Project Renewal   Renewal of a principal investigator's access to controlled-access data sets for a prior-approved project before the expiration date of a Data Access Request or Project Request (see http://www.ncbi.nlm.nih.gov/books/NBK63627/#DArequest.the_research_related_to_my_dbg).

Requester   The home institution or organization of the principal investigator that applies to dbGaP for access to NIH genomic data.

Senior/Key Persons   Collaborators at the home institution of the data submitter or requester, such as the Information Technology director.

Appendix G: TCGA Analysis Working Group Charter

PURPOSE

The purpose of The Cancer Genome Atlas (TCGA) Disease Working Groups (DWGs) is to provide clinical and cancer biology expertise for specific tumor types and to enhance tissue accrual capability. The purpose of TCGA Analysis Working Groups (AWGs) is to provide analysis of clinical and genomic data and contribute to TCGA publications on specific tumor types following sample accrual.

ORGANIZATION AND FUNCTION

DWG/AWGs are unofficial subcommittees of the TCGA Steering Committee. Participation in the DWG/AWGs is voluntary and uncompensated. Each working group will led by two cochairs, one expert from the biomedical community and one Principal Investigator from the TCGA Steering Committee. In addition, each WG will be assigned a TCGA Program Office staff member who will act as a facilitator and to communicate project needs to the WG. Coordination of all WG activity will be through a WG Coordinator from the TCGA Program Office. The WG Coordinator will provide logistic support to the DWGs and will be responsible for writing and distributing TCGA policy documents to the WGs. These groups will generally meet twice a month when active unless otherwise noted.

It is TCGA's goal to form all DWGs at the beginning of the project. DWGs will continue to function until the TCGA project has completed the collection of all data for the tumor type and completed analysis. After the DWG goals are completed, the group will suspend regular meetings during sample accrual and meet on an as-needed basis. The roles of the DWG include, but are not limited to:

1. identification of additional Tissue Source Sites for TCGA and help develop strategies for obtaining additional samples if needed;
2. provision of scientific input to the Steering Committee on patient inclusion/exclusion criteria, the type of tumor (eg, histological subtypes), and identify required and optional clinical data elements to be collected;

3. recommendation of pathology review criteria and development of biomedical contexts and clinical questions for analyses; and
4. working with bioinformatic and data analysis components of TCGA, providing expert biomedical knowledge to enhance analytical outcomes of TCGA data.

In general, TCGA assembles an AWG once $\sim 200+$ qualified cases of the specified tumor type are accrued and analysis on TCGA-funded platforms is available on those cases. The roles of the AWG include, but are not limited to:

1. working with the data generating centers to annotate/analyze publicly available data on the following TCGA-funded platforms—whole-exome sequencing, mRNAseq, miRNAseq, methylation, SNP 6 copy number, and basic clinical data;
2. determination of when a data set should be frozen for formal analyses for each TCGA manuscript and provide integrated analysis between TCGA platforms;
3. preparing a TCGA marker paper for the first set of cases in order to release data to the community within the publication moratorium set by TCGA Publication Policies.
4. organizing tumor type-specific Analysis Face-to-Face workshops ("Jamborees") once data are generated and manuscript preparation has begun; and
5. designating a data coordinator, analysis coordinator, and manuscript coordinator within the team to facilitate development of the publication.

If a TCGA AWG chooses, a smaller writing group may convene after sufficient analysis is completed. This writing group may meet more frequently and will concentrate efforts on manuscript development. The Program Office will continue to support AWG meetings and other efforts until the acceptance of the marker paper in a scientific journal. AWG meetings and other activity will not be facilitated by the Program Office after the marker paper is published; however, members are encouraged to publish additional satellite manuscripts on their own.

MEMBERSHIP

Members: Membership in a TCGA Working Group is based on the ability to provide samples to TCGA and on participation/contribution in/to the group's activity. Without exclusion, anyone who contributes samples or analysis to TCGA is eligible to be a WG member. Criteria for participation in TCGA are similar to standard criteria required for inclusion for authorship in most journals. There are no minimum number of samples required for membership.

1. Cochairs: Cochairs will be selected from the pool of DWG members. Just as participation in the DWGs is completely voluntary, being a cochair is completely voluntary. The TCGA staff member assigned to each DWG will facilitate identifying the cochairs. Once candidate cochairs are identified, the TCGA staff member will provide the names of the proposed cochairs to the Steering Committee for approval. Cochairs will facilitate the working group and report to the Steering Committee on timeline, projections, and progress toward goals for completion of analysis. Cochairs will serve for the life of the DWG and subsequent AWG unless they decided to step down from the position.

2. DCC representative: Each AWG will have a designated representative from the Data Coordinating Center (DCC) to facilitate making data publicly available and set up the publication webpage.

3. Coordinators: Coordinator roles will be assigned to members of the AWG in order to lead critical functions of the working group. Multiple roles may be filled by the same person, as long as responsibilities are clearly assigned. Roles include the following:

   a. Data Coordinator—tracks the location of all data files and the content of data freezes; ensures that data are evaluated for accuracy and consistency.

   b. Analysis Coordinator—tracks the progress of individual analysis groups and often sets specific deadlines to encourage group progress; also often provides encouragement for integrative analyses. Collects and organizes interim data analysis results.

   c. Manuscript Coordinator—serves as central contact point for assembling sections for the manuscript, ensuring consistency in style and content and figures, and takes responsibility with the Tumor WG PIs to set goals and deadlines for completion of writing sections and assembly of final paper.

TCGA WGs are open to any investigator willing to abide by the criteria for participation established for the TCGA project by the National Cancer Institute and the National Human Genome Research Institute. Each WG is open to all academic, government, and private sector scientists interested in participating in an open process to facilitate the comprehensive interpretation of the cancer genome and who agree to the criteria described below:

Criteria for participation in a TCGA DWG/AWG are:

1. Each participant will describe their specific scientific and/or clinical expertise in the cancer type assigned to a particular working group.
2. Each participant will represent one of the following: (1) a tissue source site (TSS) capable of providing samples to TCGA, (2) a TCGA-funded center generating data for TCGA samples, (3) any other institution that will provide analysis of data or other significant contribution(s) for TCGA publications.
3. Each participant is expected to contribute significantly to the project, bringing his/her particular expertise to bear on accomplishing the goals of the WG in a timely manner. Participation in the WG should consist of more than general interest or the ability to provide samples to the program, but also include substantial intellectual contributions to the WG and TCGA.
4. Each participant will take part in group activities, including active participation in periodic teleconferences to discuss the project's progress, sample specifications, clinical data requirements, data analysis, and coordinating the publication of research results. Members are expected to actively contribute to the analysis involved in developing the individual manuscripts.
5. Each participant agrees that he/she will not disclose confidential information obtained from other members of the WG.
6. Additional criteria may be added upon recommendations of the TCGA Steering Committee.

An investigator who is interested in applying for membership to a TCGA Working Group should provide a description of their expertise in the cancer type or analysis platform assigned to a specific working group, as well as a statement of agreement to abide by the Criteria for Participation listed above to TCGA Program Staff. Items that should be included are the description of the clinical and/or research expertise, planned contributions to the TCGA project(s) of interest, and a description of any retrospective samples or potential for prospective samples that could be provided to TCGA.

The description will be reviewed by TCGA staff and the TCGA Steering Committee to determine whether an investigator will be accepted into a specific DWG. The participation of members will be reviewed yearly by the Project Team.

BENEFITS OF DWG/AWG PARTICIPATION

By participating as a TSS, DWG, or AWG member, investigators collaborate in a program that compounds the statistical power of any single institutions own biospecimen set by their inclusion in significantly larger data set to which all investigators have access. Collaborative opportunities include:

Authorship on the initial in-network manuscript. The project works with journals and publication databases to ensure true, individual citation.

Joint analysis and manuscript preparation meetings of interested investigators from tissue source sites working alongside the TCGA bioinformatic experts and data analysis groups to perform additional real-time in-depth analysis. Active data analysis participants are included in authorship on additional paper(s).

Creation and participation in additional self-organizing and groups for further collaborative activity based additional questions and data surfaced in analysis working groups. These groups are well positioned to apply for additional funding targeting use of TCGA data. For example, there have already been additional groups funded to leverage TCGA data sets including the *Target Discovery and Development Network* and the *caBIG In Silico Research Centers of Excellence programs*.

Participation in a project that encourages support for trainees and junior investigators, for whom the project's scientific leadership will provide letters of support naming an individual's contributions.

In addition to the above, the following is a list of specific benefits that help justify the contribution of these specimens and clinical data to TCGA:

Investigators from contributing institutions are asked to actively participate on project working groups in which are developed the specific case and specimen inclusion/exclusion criteria of the cancers to be studied.

The project will bear the significant laboratory costs of comprehensively processing and characterizing your biospecimens. For the contributing institution, resources otherwise spent on characterization and information technology are freed for more focused investigator-initiated and possibly clinical studies, deriving from hypotheses generated from the TCGA genomics data sets.

The project will comprehensively process and characterize your biospecimens across multiple genomic platforms and then standardize and integrate the profiling data and clinical annotation across all the contributing sites. The resulting data are immediately available back to your investigators. The characterization platforms are described at http://tcga.cancer.gov/wwd/program/ and include:

❑ Whole-genome (20% of cases) and whole-exome sequencing, on both tumor and germline DNA for up to 200 cases,

❑ Expression profiling, on both mRNA and microRNA,

❑ Epigenomic characterization, and

❑ Copy number and single nucleotide polymorphism analysis.

The project has developed, continued to improve, and made broadly available software for a broad range of data management and associated analysis tools. Access to data, documentation, and these tools is provided to your investigators.

NIH encourages contributing sites to retain residual material from every case profiled by TCGA, to allow for more focused biochemical or signaling pathway studies based upon hypotheses generated from the TCGA genomics data sets.

TCGA requires a clinical data set to be provided along with the samples, providing the contributing site exclusive control of the broader clinical story surrounding a particular patient. These data can be the basis for investigator-driven collaborations launched within the project analysis working groups as described above.

Index

Note: Page numbers followed by "*f*" and "*t*" refer to figures and tables, respectively.

A

Access controls, 56
Accountability tracking, 11
Algorithms, testing of, 45
Analysis coordinator, 129
Analysis Working Group (AWG), 3,
    59–62, 91
Annual scientific symposia, 26–27
Approved user, 117–122, 124
Audiences
    categories, 17–18
    concerns of, 18–20
    identification of, 17–18
    priority, 21–22
Authorship, traditional, 65
Authorship models, 65
    advantages and disadvantages
        of, 66*t*
AWG. *See* Analysis Working Group
    (AWG)

B

Barcode labelling, 35
BCR. *See* Biospecimen Core Resource
    (BCR)
Binary Sequence Alignment Map format
    (BAM), 52
Biospecimen Core Resource (BCR), 2–3,
    32, 34, 46, 61
Biospecimen source sites
    contractual obligation and payment
        plans for, 37–38
Budgetary vigilance
    by data providers, 78
    by Program Office, 78

C

Cancer Genome Atlas Research
    Network, 101
Cancer Genomics Hub (CGHub), 3, 52,
    94, 115
Cancer-associated genes, identifying, 1–2
CDEs. *See* Clinical data elements (CDEs)
Central biospecimen processing facility,
    establishing, 32–33
Central data management center, 51
Central processing facility, 32–33

Centralized data management center,
    creation of, 51–52
Centralized Institutional Review Board, 36
CGHub. *See* Cancer Genomics Hub
    (CGHub)
ChiP-seq (chromatin immunoprecipitation
    following by high-throughput
    sequencing), 43
Clinical correlation analysis, 60
Clinical data collection, management of,
    38–41
    practical considerations for, 40–41
    sharing clinical data, 40
Clinical data elements (CDEs), 40
Closure planning, 73
Closure protocols, standardizing, 74–75
Cluster of cluster analysis, 63
Clustering analysis, 62–64
CMS. *See* Content management system
    (CMS)
Cochairs, 129
Collaboration, 82
Collaborator, 26, 124
Communication, 82–83
Communications devices, 26–27
    annual scientific symposia, 26–27
    visual identity, 27
Communications strategies, 15, 64–65
    audiences and stakeholders, identifying,
        17–18
    basis of, 17
    communications devices, 26–27
        annual scientific symposia, 26–27
        visual identity, 27
    development, 15–17
        note on press releases, 16
        sharing messages, 16–17
        step 1:challenges, 18–19
        step 2:consideration, 19–20
        step 3:messaging, 20
        step 4:tactics, 21
        step 5:priority, 21–22
        step 6:tactic evaluation, 22
        step 7:strategy evaluation, 22
    project and policy changes, 23
Consent protocols, establishing, 35–36
Consortium and individual authors, 65
Consortium authorship, 65
Content management system (CMS), 57

Contractual obligation, 37–38
Contributing investigators, 116, 124
Controlled-access environment, 56
Coordinators, 127, 129
Copy number data, 44, 44*f*

D

DAR. *See* Data Access Request (DAR)
Data access process, 122
Data Access Request (DAR), 116, 124
Data analysis, 59
    analysis structure and methodology,
        62–64
        emerging analytical tools, 63
        outliers/exceptional cases, 63–64
        study design, 62–63
    analysis teams, establishment of, 61–62
        AWGs and diversity of expertise, 61
        contributions, managing, 61–62
    efforts, 5, 13
    practical considerations, 64–66
        aiming journals, 64–65
        authorship models, 65
        falling into formula, 64
        timeliness versus scientific merit,
            65–66
    preconceived questions, 59–61
Data collection forms, 38, 40
Data Coordination Center (DCC), 3, 52,
    61, 77, 94
    data versioning, 54
    representative, 129
    role of, 51–52
Data coordinator, 129
Data Derivative, 124
Data file identifier, 54
Data generation, 4, 13, 43, 70
    data generation model, building, 43–46
        data types, 43–45
        multicenter data generation, 46
        single-centerdata generation, 46
        technologies and methods
            employed, 45
    pipeline, establishing, 47–48
    proper tracking of, 48
    quality control measures, 47–48
Data generation center(s), 32–33
    sample distribution to, 34–35

Data management center, centralized, 51–52

Data management system, 40–41, 56

Data providers
  budgetary vigilance by, 78
  standardized closure protocols across, 74–75

Data security checks, 55

Data storage and dissemination, 4, 13, 51, 70
  appropriate security and access controls, 56
  centralized data management center, creation of, 51–52
  collect, store, and version data and metadata from various sources, 53–54
  data and metadata tailored to diverse project stakeholders and end users, redistribution of, 56–57
  quality control measures for submitted data, implementation of, 55
  standard data and metadata formats, 52–53

Data types, 43–45, 44f

Data Use Agreement, 35

Data Use Certification (DUC) agreement, 115, 124
  statement of policy, 115–116
  terms of access, 116–123
    data security and data release reporting, 119–120
    intellectual property, 121
    non-endorsement, indemnification, 122
    non-identification, 118
    non-transferability, 118–119
    public posting of approved users' research use statement, 118
    requester and approved user responsibilities, 117
    research dissemination and acknowledgment of NIH genomic study data sets, 121
    research use, 116
    research use reporting, 122
    termination and violations, 123
  The Cancer Genome Atlas (TCGA), 115

Data versioning system, 54

Database of Genotypes and Phenotypes (dbGaP), 94, 115

dbGaP Approved User Code of Conduct, 124

dbGaP Security Best Practices, 119–120

dbGaP. *See* Database of Genotypes and Phenotypes (dbGaP)

DCC. *See* Data Coordination Center (DCC)

Decision Trees, 75

Disease Working Groups (DWGs), 91–92, 127

DNA methylation, 44, 44f

DNA sequencing, 44, 44f

Documentation
  of IRB approval, 36
  for project closure, 76–77
  user documentation, 57

DWG/AWG participation, 127
  benefits of, 131–132
  criteria for, 130

E

ENCODE project, 43

Endpoint of a project, 73

Ethical management of samples, information, and derived data sets, 70–71

External Scientific Committee, 82–83

F

FASTQ format, 52

Federal Information Management Security Act (FISMA), 56

Federally funded studies, 73–77

File format checks, 55

FISMA. *See* Federal Information Management Security Act (FISMA)

Flexibility, 81

Freedom of Information Act (FOIA), 82

Freedom-to-publish criteria, 98

G

Gantt charts, 75

GCCs. *See* Genome Characterization Centers (GCCs)

GDACs. *See* Genome Data Analysis Centers (GDACs)

Genome Characterization Centers (GCCs), 2–3

Genome Data Analysis Centers (GDACs), 3

Genome Sequencing Centers (GSCs), 3
  closure tasks for, 73–77

Genome-Wide Association Studies (GWAS), 62–63, 115

Genomic data generation, 76

Genomic platforms, 62–63, 81

Genomic regions, identifying, 1–2

Genomic sequencing, 1, 52

Global analysis publication, 97, 100–101

Government officials and policy makers, 18

Graphical user interface (GUI), 56–57

GSCs. *See* Genome Sequencing Centers (GSCs)

GUI. *See* Graphical user interface (GUI)

GWAS. *See* Genome-Wide Association Studies (GWAS)

H

Health care providers, 18

Histone modification, 43

HotNet, 63

Human subjects research, 118

Human-readable sample identifier, 35

I

ICGC. *See* International Cancer Genomics Consortium (ICGC)

Identity checks, 55

Industry as potential audience, 18

Information Technology (IT) Director, 116, 124

Institutional Certification, 115, 124

Institutional Review Board (IRB), 35–36, 116
  centralized, 36
  closure, 76–77

Institutional-level closure, 73–77

Intellectual property (IP), 121

International Cancer Genomics Consortium (ICGC), 25, 82

Interpreted data, 44–45

Investigation Design Format (IDF) file, 53

IP. *See* Intellectual property (IP)

IRB. *See* Institutional Review Board (IRB)

L

Laboratory Information Management System (LIMS), 40–41, 48

Large-scale genomics data set, 59

Large-scale genomics research project, 3–4, 9–12, 31, 51–52, 64–65, 69, 73, 81
  milestones for, 10t

LIMS. *See* Laboratory Information Management System (LIMS)
Logo. *See* Visual identity

M

MAGE-TAB format, 76, 111
Manuscript coordinator, 129
Material transfer agreement (MTA), 36
Media, 18
Messages, 20, 22
 priority, 21–22
 selection of medium, 21
 sharing, 16–17
Milestones for large-scale genomics research projects, 9–10, 10*t*
miRNA sequencing, 44, 44*f*
Molecular data, types of, 44
Moratoriums, 65–66
mRNA sequencing, 44*f*
MTA. *See* Material transfer agreement (MTA)
Multicenter data generation, 45–46
Mutation Annotation Format (MAF), 52, 103

N

National Cancer Institute (NCI), 1–2, 100
National Human Genome Research Institute (NHGRI), 1–2
National Institutes of Health (NIH), 56, 115, 117, 121–122, 132
 genomic data, 116, 118
 Genomic Study Data Sets, acknowledgment of, 121
NCI. *See* National Cancer Institute (NCI)
NHGRI. *See* National Human Genome Research Institute (NHGRI)
NIH. *See* National Institutes of Health (NIH)

O

Outliers, 63–64

P

Patient descriptors, 31–32
Patients and advocacy organizations, 18
Payment plans for biospecimen source sites, 37–38
PI. *See* Principal Investigator (PI)

Pipeline activities, examples of, 13–14
 data analysis efforts, 13
 data generation, 13
 data storage and dissemination, 13
 quality control, auditing, and reporting, 14
 sample acquisition and enrolment, 13
Pipeline of project, designing, 10–14
 pipeline diagram, building, 11
 project activities, establishing, 10–11
 relationships between activities, establishing, 11
 TCGA workflow diagram, 12*f*
Policy considerations, 4–5
 data analysis efforts, 5
 data generation, 4
 data storage and dissemination, 4
 quality control, auditing, and reporting, 5
 sample acquisition and characterization, 4
Preconceived questions, 59–61
Predata generation steps, 32–33
Press releases, 16
Principal Investigator (PI), 116–117
Pro-bono data providers, 82–83
Processed data, 44
Program Office, 73, 77–78, 82
Program-level closure, 77
Project activities
 establishment of, 10–11
 relationships between, 11
Project and policy changes, 23
Project closure, 73
 budgetary considerations, 77–78
 budgetary vigilance by data providers, 78
 budgetary vigilance by Program Office, 78
 documentation, 76–77
 levels of, 73–77
 institutional level, 73–77
 program level, 77
 publication level, 73–77
 workflow and timelines for, 75
Project stakeholders, 44–45, 56–57
Prospective sample collection, 34
Public as potential audience, 18
Publication guidelines, 97
Publication-level closure, 73
Purpose of the project, 7–8

Q

QC check, 54
Quality control, auditing, and reporting, 5, 14, 69
 ethical management of samples, information, and derived data sets, 70–71
 providing quality reports to stakeholders, 71
 quality management issue, example of, 71–72
 quality metrics, establishment of, 69–70
Quality control measures
 establishing data generation pipeline and, 47–48
 for submitted data, 55
Quality management system, 69–72
Quality metrics, establishment of, 69–70

R

Raw data, 44, 47–48
Report on quality issues, 71
Reporters, 16
Requirements, gathering, 7
 key stakeholders, identifying, 9
 pipeline, designing, 10–14
 pipeline activities, examples of, 13–14
 project milestones, setting, 9–10
 purpose of project, 7–8
Restricted access, 56
Retrieval system, 57
Retrospective sample collection, 34
RNA sequence, 44

S

Sample acquisition, 31, 59, 69–70
 central biospecimen processing facility, establishing, 32–33
 centralized IRB, 36
 and characterization, 4
 clinical data collection, management of, 38–41
 practical considerations for, 40–41
 sharing clinical data, 40
 consent protocols, establishing, 35–36
 contractual obligation and payment plans for biospecimen source sites, 37–38
 individual IRB approval rulings, 36
 potential TSSs, identifying, 37

Sample acquisition (*Continued*)
   sample processing and distribution to
      data generation centers, 34–35
   sample qualification metrics,
      establishing, 33–34
   sample set, defining, 31–32
Sample and Data Relationship Format
   (SDRF) file, 53
Sample qualification, 33–34, 39*f*
Sample size, minimizing, 32
Scientific merit, timeliness versus, 65–66
Scientists as potential audience,
   17–18, 23
Security Best Practices Requirements, 118
Sequence Alignment Map (SAM) format,
   52
Single-center data generation, 45–48
   data generation, 47
   data submission, 47–48
   sample intake, 47
   sample preparation/nucleic acid
      extraction, 47
SNP arrays, 63
Stakeholders, 72, 82–83
   identifying, 9, 17–18
   providing quality reports to
      stakeholders, 71
Standard data and metadata formats,
   52–53
Strategy evaluation, 22
   project and policy changes in, 23
Study completion and closure, 74

T

Tactics, 16–17, 21
   evaluation, 22
TCGA. *See* The Cancer Genome Atlas
   (TCGA)
The Cancer Genome Atlas (TCGA), 1–3,
   27, 37–38, 44, 73, 97, 115
   data levels and data types, 44*f*
   DCC submission workflow, 54*f*
   external stakeholders, 82–83
   funds, 81
   Genome Sequencing Centers, 74–75
   internal stakeholders for, 82–83
   Melanoma Analysis Working Group, 59
   Program Office, 61–62, 77–78
   project analysis workflow, 60*f*
   publication moratoriums, 65–66
   Research Network, 2, 100
   sample qualification metrics from, 33–34
   tissue providers, 75
   TCGA Analysis Working Group (AWG)
      Charter, 127
      benefits of DWG/AWG participation,
         131–132
      membership, 128–130
      organization and function, 127–128
      purpose, 127
   TCGA data, 44*f*
      in publications and presentation,
         100–101
      for research purposes other than
         publication and presentation, 101

tumor analysis projects, 59
   visual identity, 27
   workflow diagram, 12, 12*f*, 91
1000 Genomes Project, 7–9, 43
Tissue provider contracts, 92
Tissue providers, 75, 78
Tissue source sites (TSSs), 32–33, 38,
   40–41, 92
   identification of, 37
Traditional authorship, 65
Transparency, 81–82
TSSs. *See* Tissue source sites (TSSs)

U

Unique identification number, 71
User documentation, 57

V

Validators, 55
Verbal agreement, 64–65
Visual identity, 27

W

Web Application Programming
   Interface, 57
Whole-exome sequencing data, 46

Y

YouTube, 17

Printed in the United States
By Bookmasters